INSTANT GRATIFICATION

–21st Century Art & the Mobile Phone Camera–

CRISTOPHER LAPP

Instant Gratification, 21st Century Art & the Mobile Phone Camera

Copyright © 2010 Cristopher Lapp

All rights reserved. No part of this book may be reproduced or transmitted in any form or
by any means, electronic or mechanical, without written permission of the author,
except where permitted by law. To request permission and all other inquiries,
contact West Hollywood Books at info@instantgratificationbook.com

Published 2010 by West Hollywood Books™

Library of Congress Cataloguing in Publication Data
Lapp, Cristopher.
Instant Gratification, 21st Century Art & the Mobile Phone Camera
2010938355

ISBN: 978-0-615-40959-7

Printed in The United States of America by BookMasters, Inc.
October 2010.
30 Amberwood Parkway Ashland, OH 44805
#M7919

Cover and Interior Design by Mollie Crutcher
Art consultation by Justin Gilanyi

10 9 8 7 6 5 4 3 2

First Edition

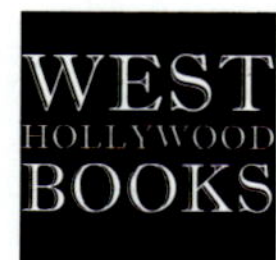

CONTENTS

FOREWORD

The newest member of my family, Mason William Ziebarth, was born earlier this week. In less than an hour I was sharing an image of his miraculously scrunchy face. There was no text, no identifying caption or misspelled adjectives, no modifiers, no note; just a perfectly normal image of a perfectly extraordinary baby. The image WAS the message.

As consumers of images, with our current understanding of what immediacy entails, the marketing claims of Kodak's Instamatic can now be seen as almost laughable with its lures of immediacy and ease of use; the Polaroid revolution gave us the instant gratification we craved but, ultimately, it was an object that was mechanically presented from the camera's gaping slot—an object that carried an image, but an object none-the-less. The evolving and free-flowing transmission of images that comes with our mobile phones has unburdened us from the object-orientation of image making, and is changing the options we have in our approach to our lives and how we picture them. Our phones have become not only the means by which we create images, but the vehicles by which we receive them, store them, and share them. Images are evolving beyond their status as commodity to become the actual currency of experience.

The Photography community has been wallowing in the idea of visual literacy for decades. We have mobilized pedagogical strategies, grant-driven initiatives, and educational outreaches not merely to attain this goal, but to define it. Ironically, the constant shimmer of images that inserts itself into our lives via our mobile phones is the best hope yet of the visual literacy ideal becoming the commonplace—not programmatically, but organically.

We regularly tote in pockets, briefcases, and hanging from belt loops, not simply a new way of making images, but a new way of thinking pictorially; a way that cannot (and should not) be straight-jacketed into any traditional notion of capture, transmission, and reception that is tethered to the creation of objects. While we are not as of yet seeing the emergence of a new vocabulary of images—our understanding of how images should be made is learned, after all—we are, at the very least, beginning to assimilate a new syntax of images; a new grammar of pictography.

Are we in the midst of an image making revolution? You bet. It is not a revolution that will supplant established modes, however, but one that redefines and extends them. Is it art? Perhaps. I hope that one day Mason William Ziebarth and a new generation of artists will shepherd me into a new mind-set that will allow me to recognize that potential.

Tim B. Wride

Los Angeles

Tim B. Wride is the founding Executive Director of The No-Strings Foundation, a non-profit foundation established in 2004 whose mission is to provide direct funding to photographic artists.

Mr. Wride was previously employed at the Los Angeles County Museum of Art (LACMA) as the Curator and Interim Head of the Department of Photographs. During his 14-year tenure as Curator of Photography at the LACMA (1992-2006), Wride curated over twenty-five permanent collection focus exhibitions as well as numerous larger exhibitions including "Shifting Tides: Cuban Photography after the Revolution" that was the first survey of the work in the U.S.

He is the author of numerous catalogues that accompanied his exhibitions, and also contributed the photography component and an anthology essay to the exhibition "Made in California: Art, Image, and Identity, 1900-2000" (2000), he co-curated and wrote the Aperture monograph for "Pirkle Jones: Sixty years of Photography" (2001) a travelling exhibition that premiered at the Santa Barbara Museum of Art; and also curated "To Protect and To Serve: Photography from the LAPD Archives" (2002) that has traveled internationally.

INTRODUCTION

The mobile phone camera, a multipurpose and internationally used tool, is bringing the creation of art to the masses. This book was made to show the relationship between privacy, individualism, and the temporal nature of art in the digital age through the convenience of the mobile phone camera. Most digital images never touch paper. They strictly exist through technology and electronic energy. The transient characteristics of these images directly confront our traditional perceptions and consciousness of what art and photography are, how they are made, and mostly how it is shared. The instant gratification provided by the image itself trumps the importance of image quality. The simple process of clicking a button on a mobile phone and making a photographic-like image that is public or private, legal or illegal, by a small device that over two billion people worldwide now use, is a juggernaut of change to society and the artistic world.

During the past six years of this project, I have made more than 26,000 images around the world using the tiny digital cameras built into several different mobile phones. Randomly capturing all situations and experiences of life was my primary goal—simply replicating what is happening billions of times each day. The ease in capturing any moment struck me as being a very powerful agent of change in human history. Boundaries of privacy as we currently understand them are quickly disappearing and will perhaps be gone within a generation.

A similarly significant technological advance was the moveable type printing press developed by Johannes Gutenberg[1] in the mid-1400s. No longer did the elite control information. Knowledge could be distributed to the general public rapidly and much more affordably. Today we're experiencing a similar phenomenon with the mobile phone camera. In the digital age, instant electronic imagery has quickly become something the entire world has access to. Within a matter of seconds, an image can be captured and delivered worldwide. The mobile phone camera is bringing the opportunity to create a new form of instant art to the masses. The public is quickly gaining the confidence to make and share personal moments and images with anyone and everyone, physically or electronically. These images are completely disposable, rarely to be viewed twice. Internet sites like YouTube[2], Facebook[3] or the tens of thousands of chat and social websites clearly demonstrate this point. Whether capturing a friend's birthday party for a quick laugh, an instant love connection, or an internationally significant disaster, mobile phone cameras are everywhere and people are using them.

In addition to sharing some of the images I've captured on my journeys, the other goal of this book is to visually demonstrate that everyone with a mobile phone camera can create "art," instant electronic "paintings." With the camera phone technology in its infancy, the images being produced are more reminiscent of the visual Impressionist style and emotional quality of a Van Gogh, Monet, Cassatt, or Seurat painting.

Once again, the instant emotional reaction caused by the image itself trumps the importance of image quality, avoiding the delayed gratification characteristic of more traditional photography. We are using our mobile phone cameras to document our hectic daily lives. This utilitarian tool and its camera is perhaps our only way to record of a life completely bombarded with visual information. These images somehow justify our own value in the world, whether the world is interested or not.

The future will obviously bring more advanced technology and clearer images. But it is my belief that these images, for better or for worse, will remain fleeting by-products of a world society being taken over by powerful technology and defined by the desire for instant gratification.

Cristopher Lapp

---

[1] Wikipedia. Johannes Gensfleisch zur Laden zum Gutenberg (c. 1398 – c. February 3, 1468), German.
[2] Wikipedia. YouTube is a web service created in 2006. By January 2008 alone, nearly 79 million users had made over 3 billion video views.
[3] Wikipedia. Facebook has currently an estimated 500,000,000 users worldwide as of July 2010.

MY FIRST MOBILE PHONE IMAGES

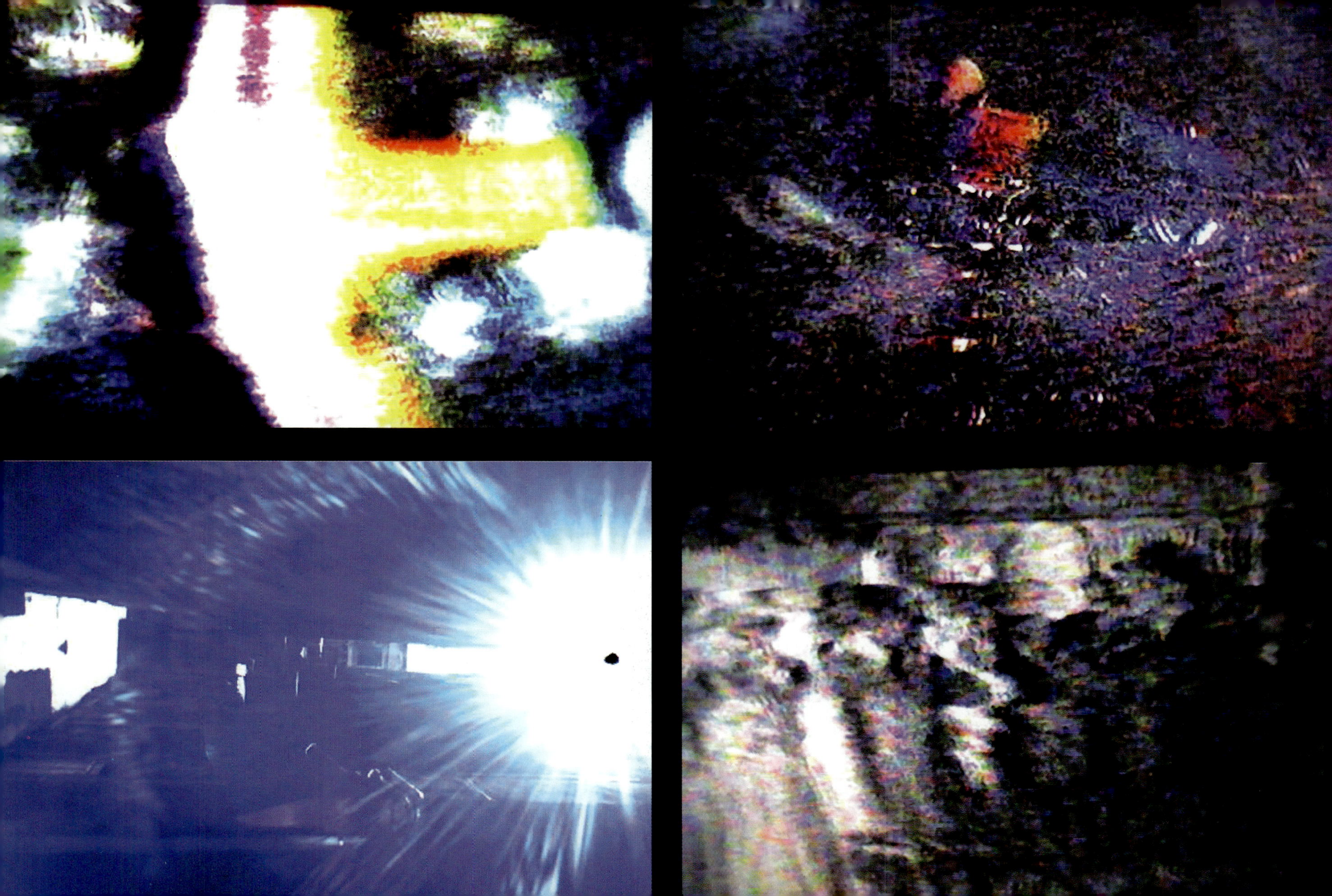

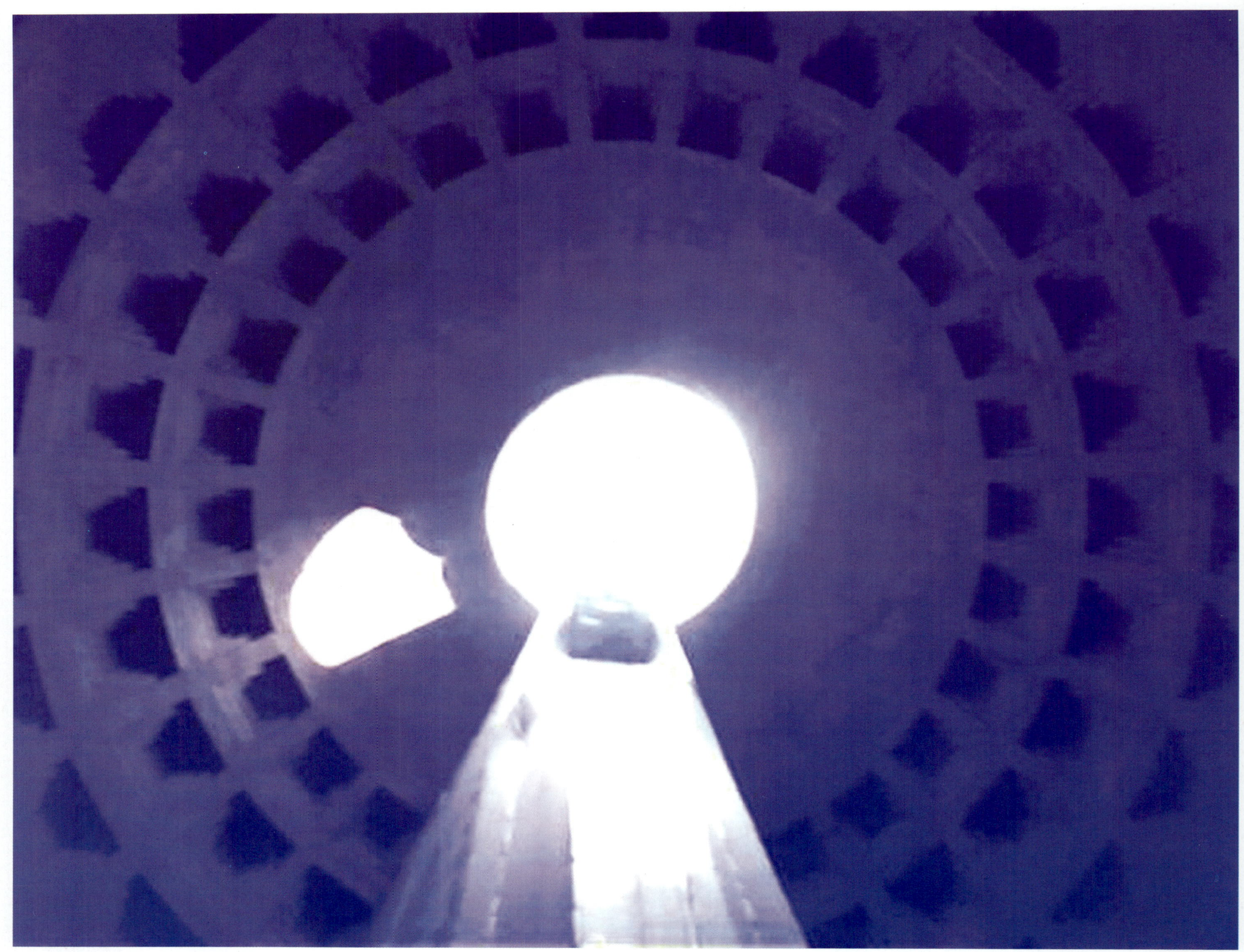

3 My First Mobile Phone Images

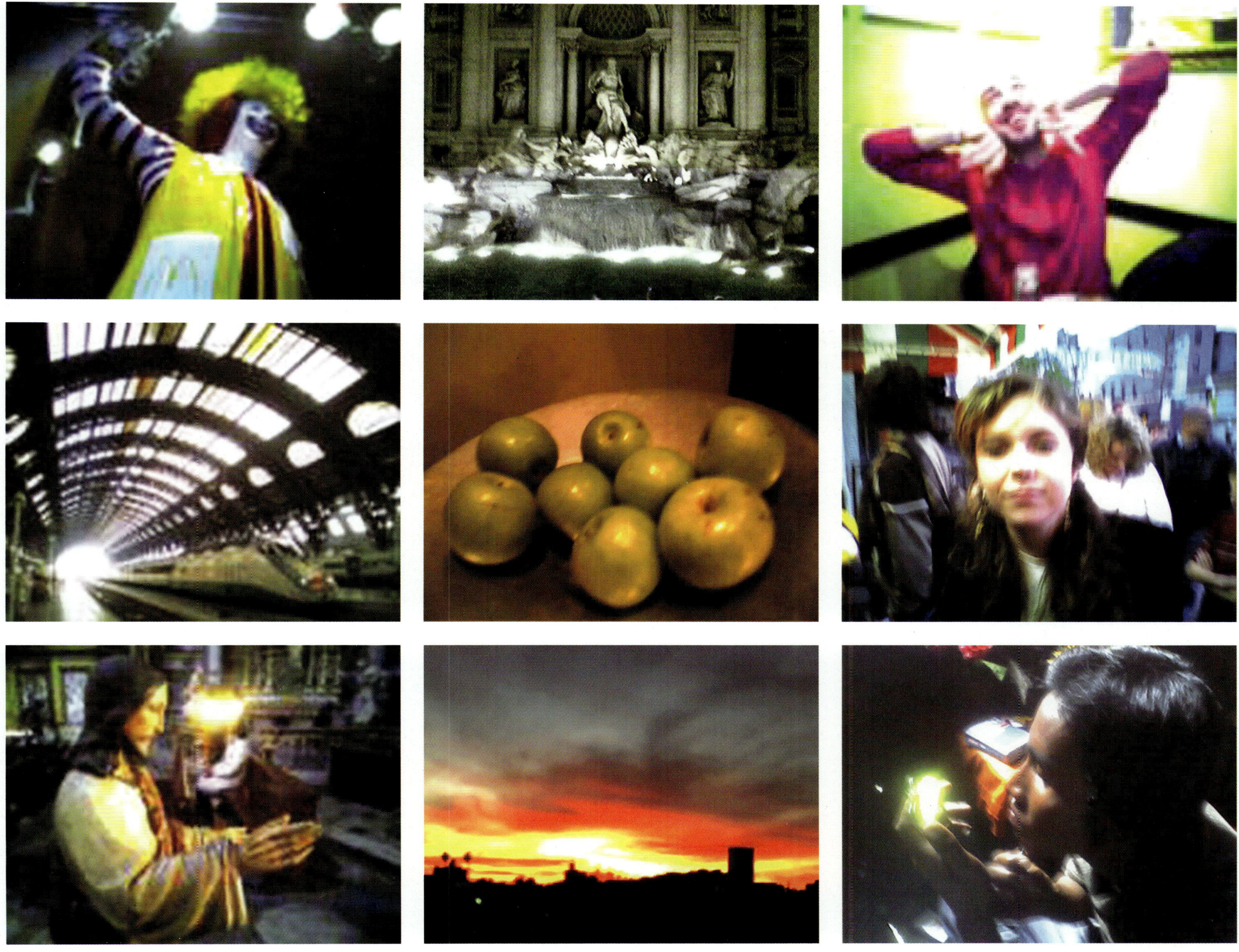

COMMUNICATION

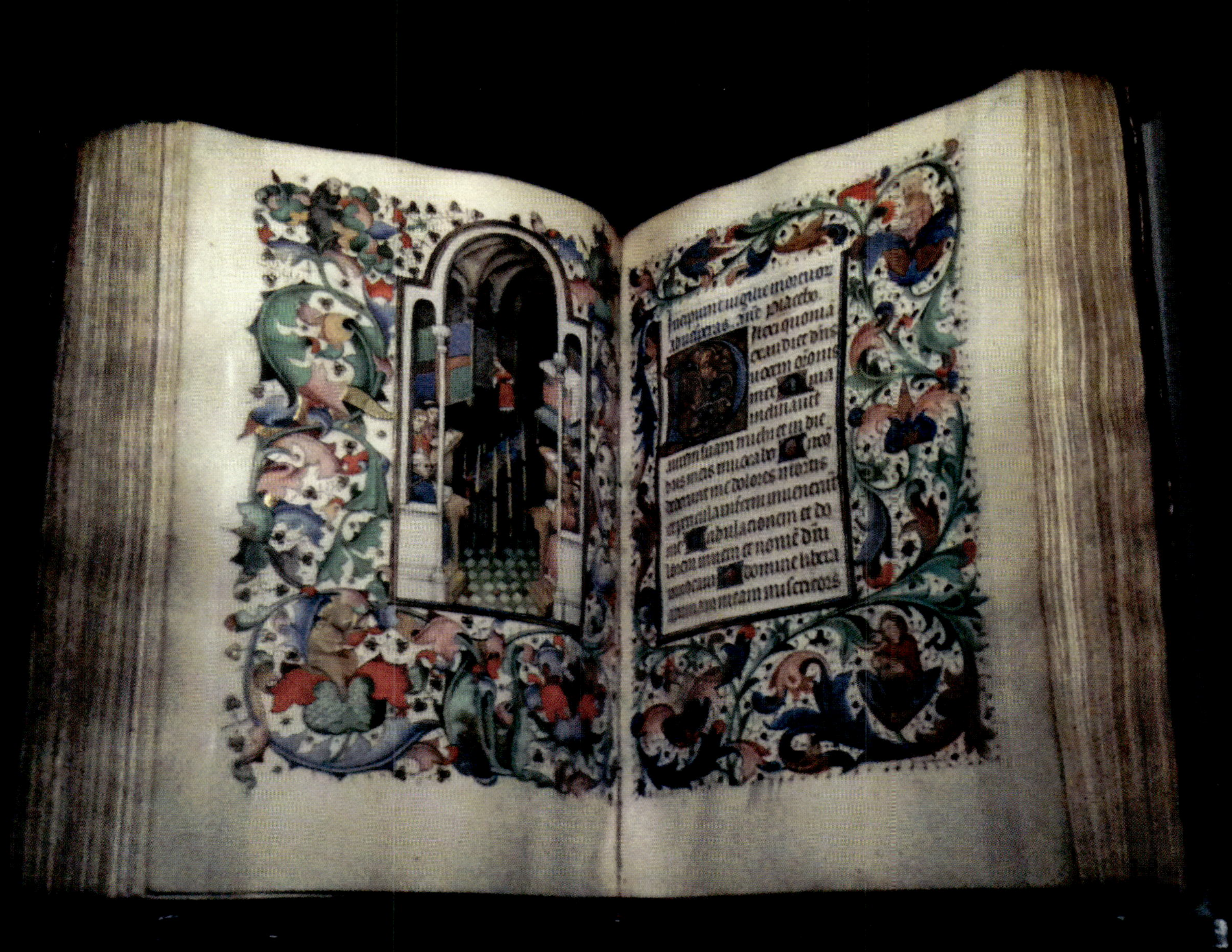

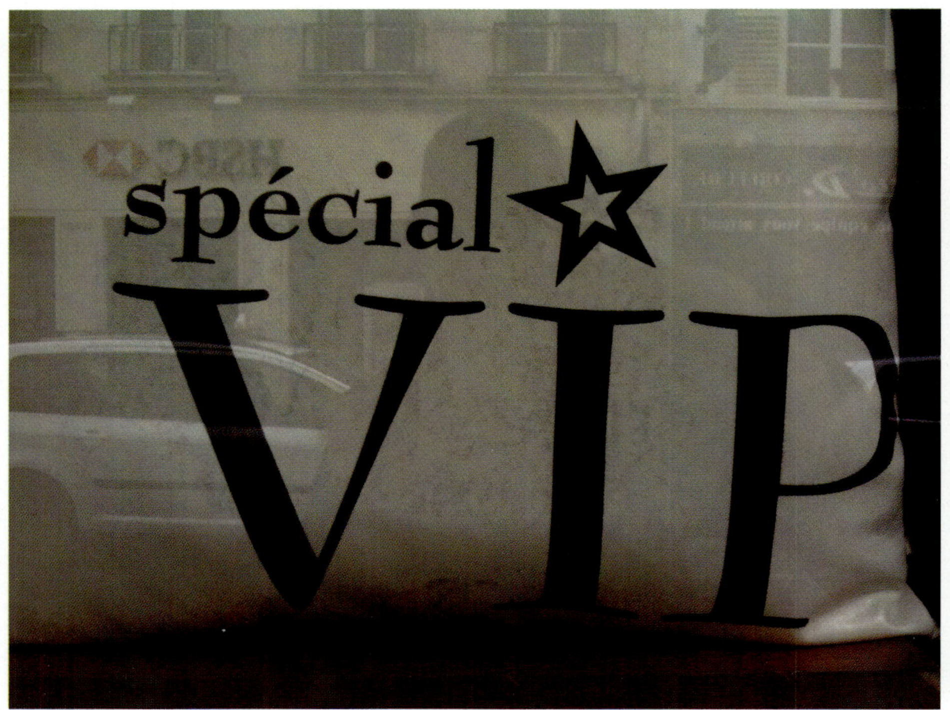
spécial
VIP

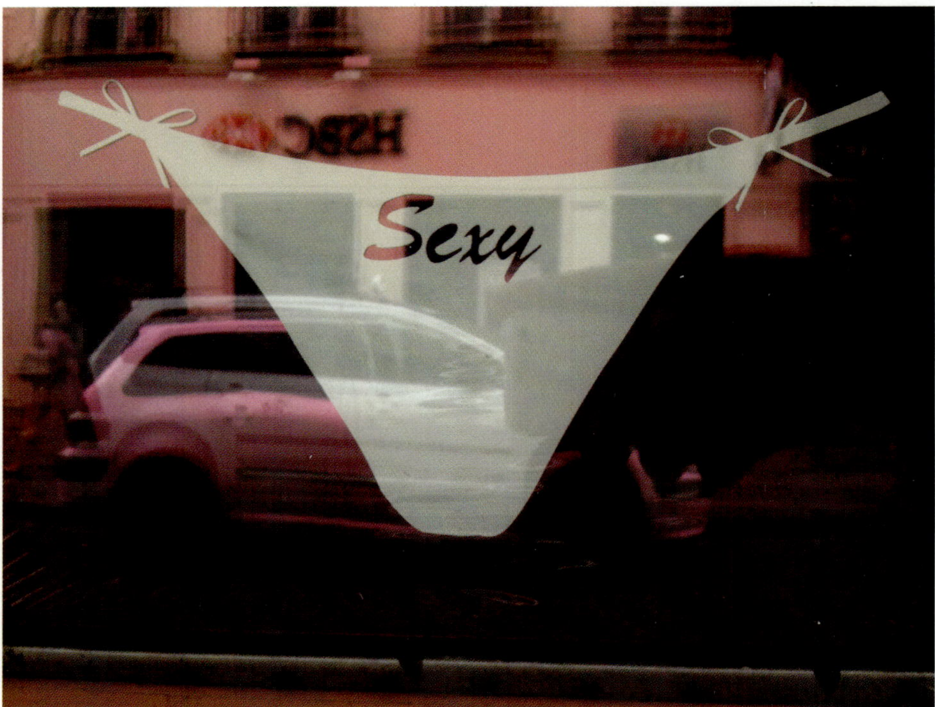
Sexy

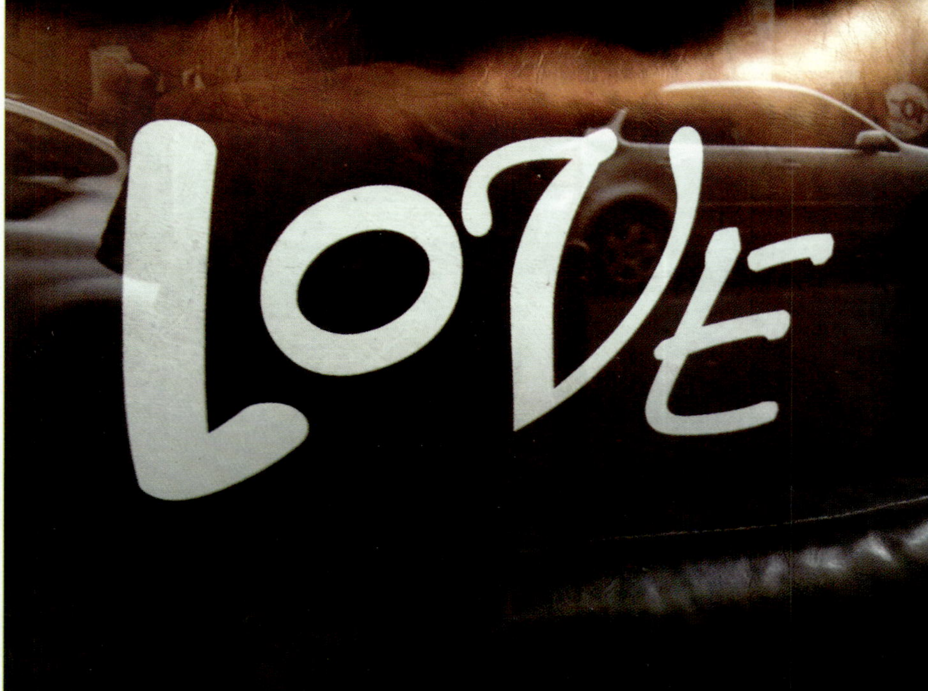
LOVE

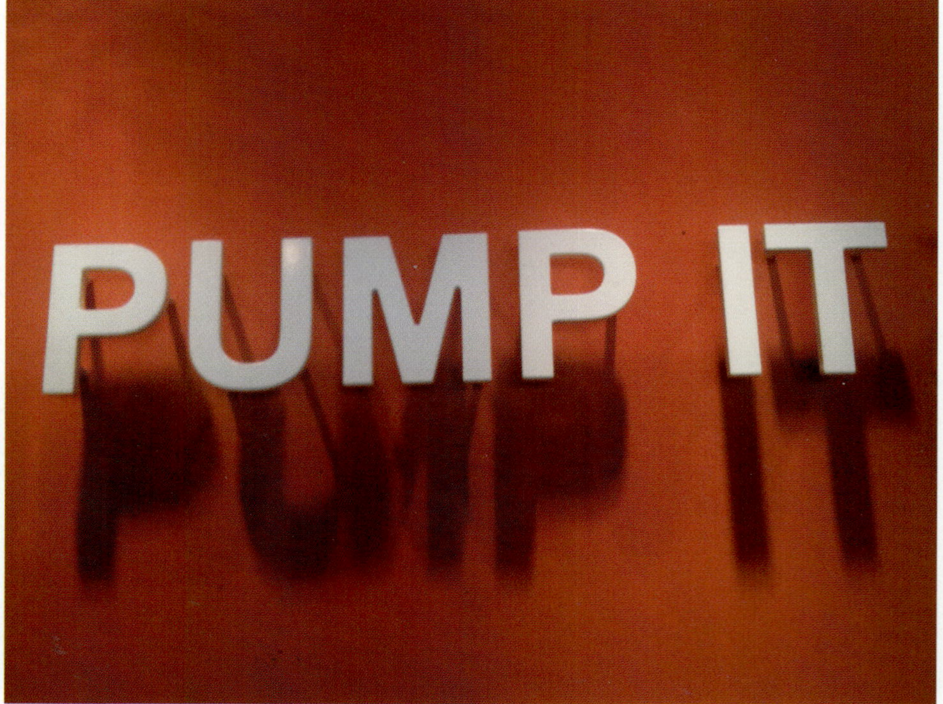
PUMP IT

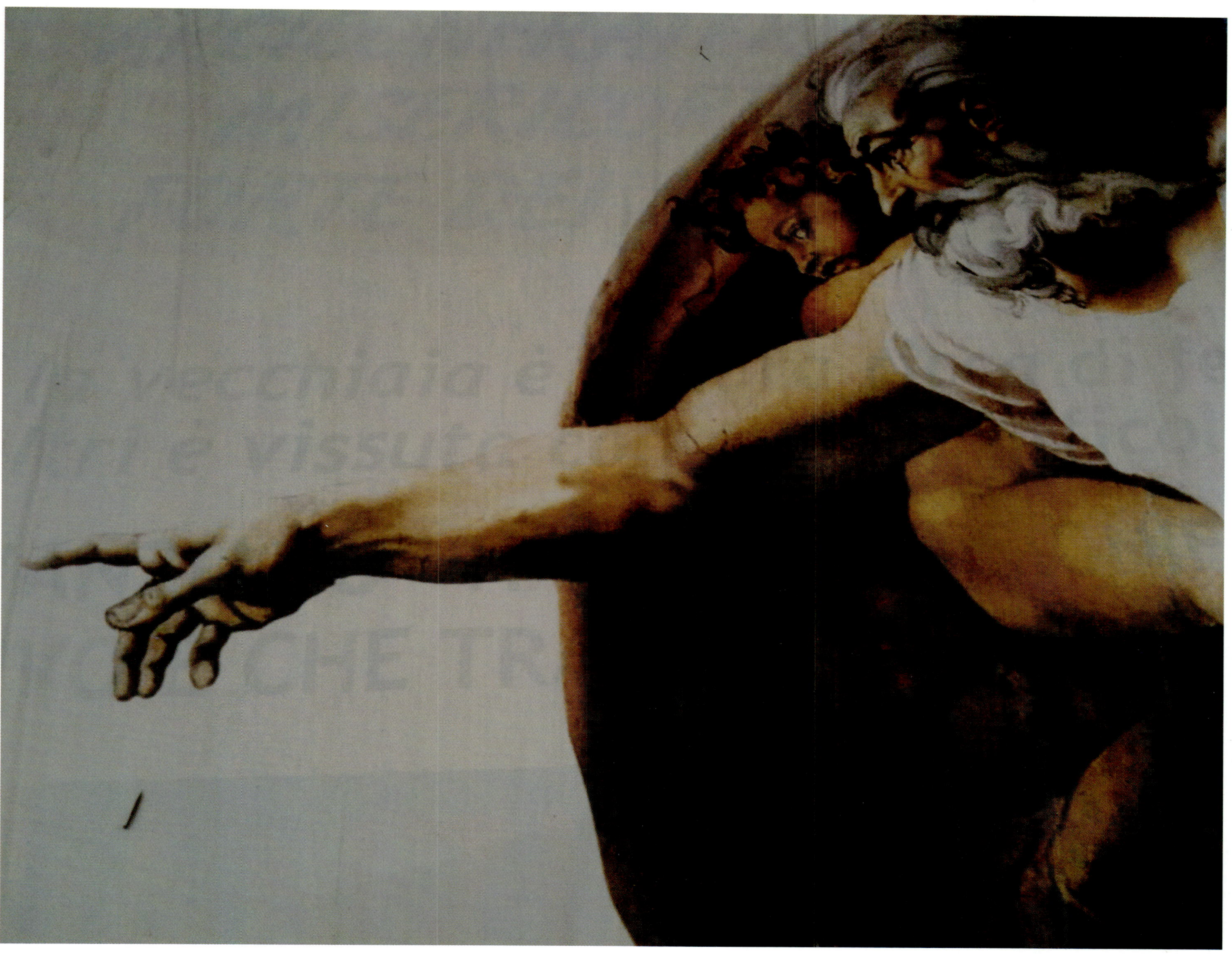

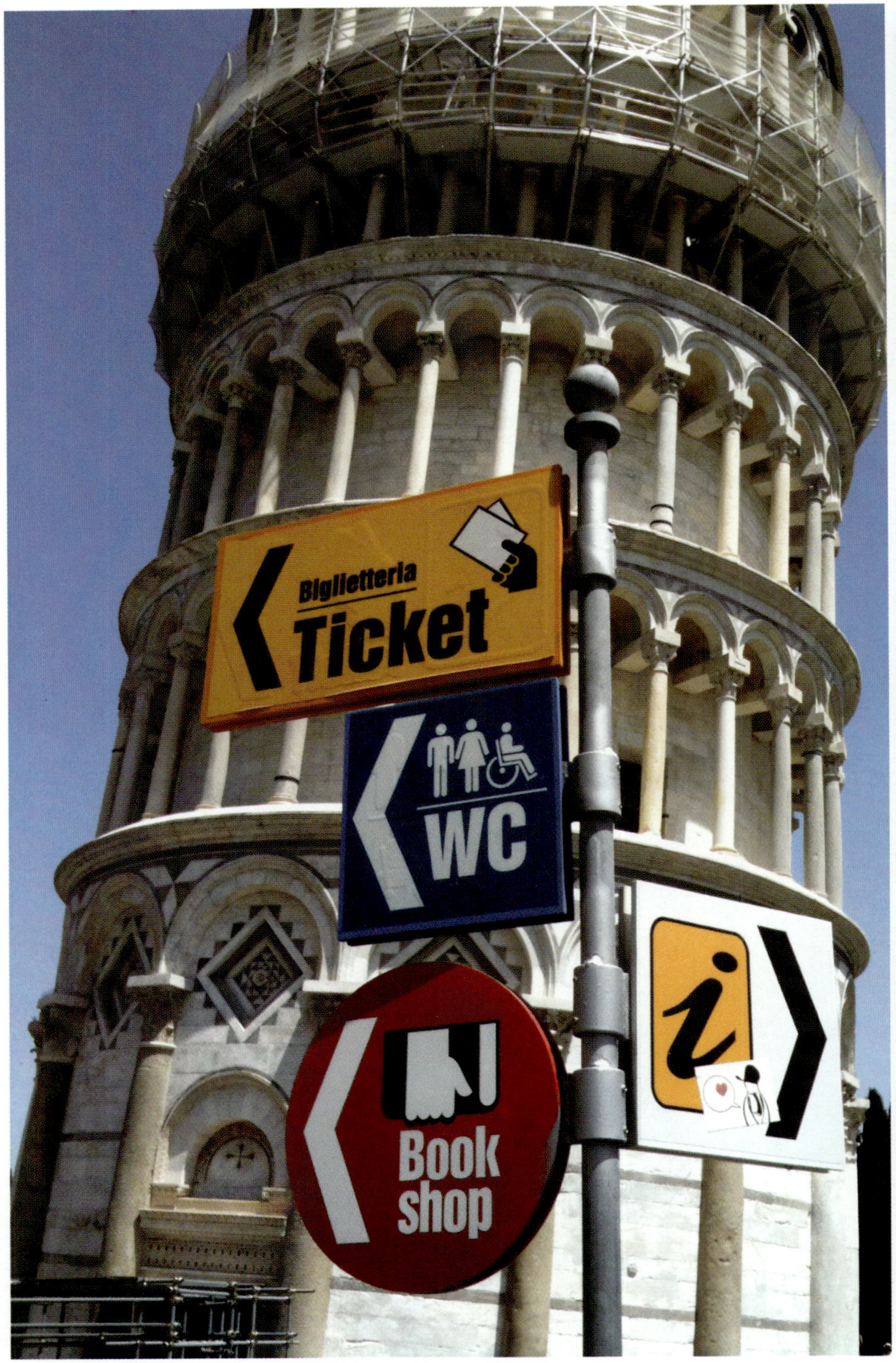

Biglietteria
Ticket
WC
Book
shop
i

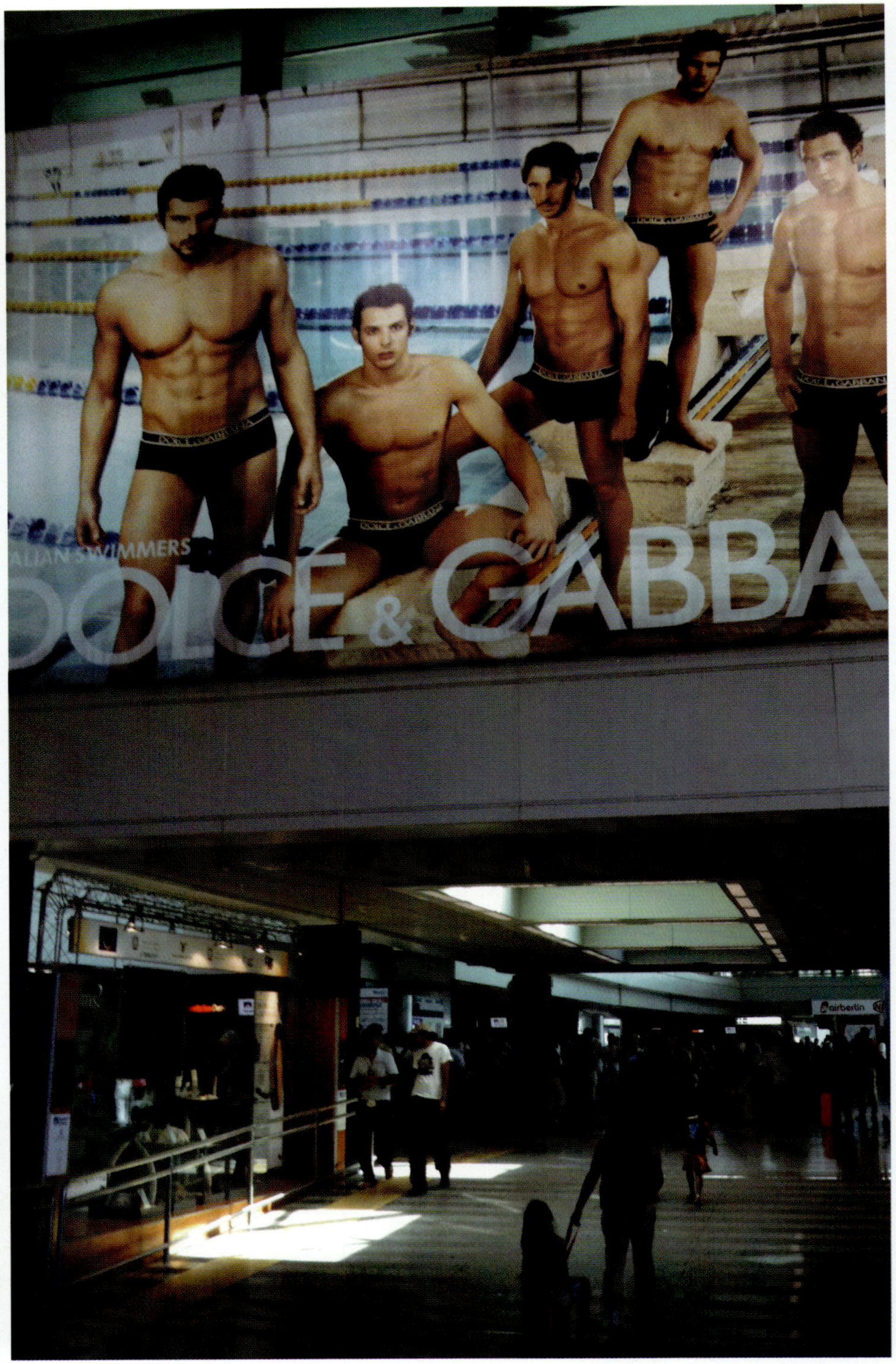

ITALIAN SWIMMERS
DOLCE & GABBA
airberlin

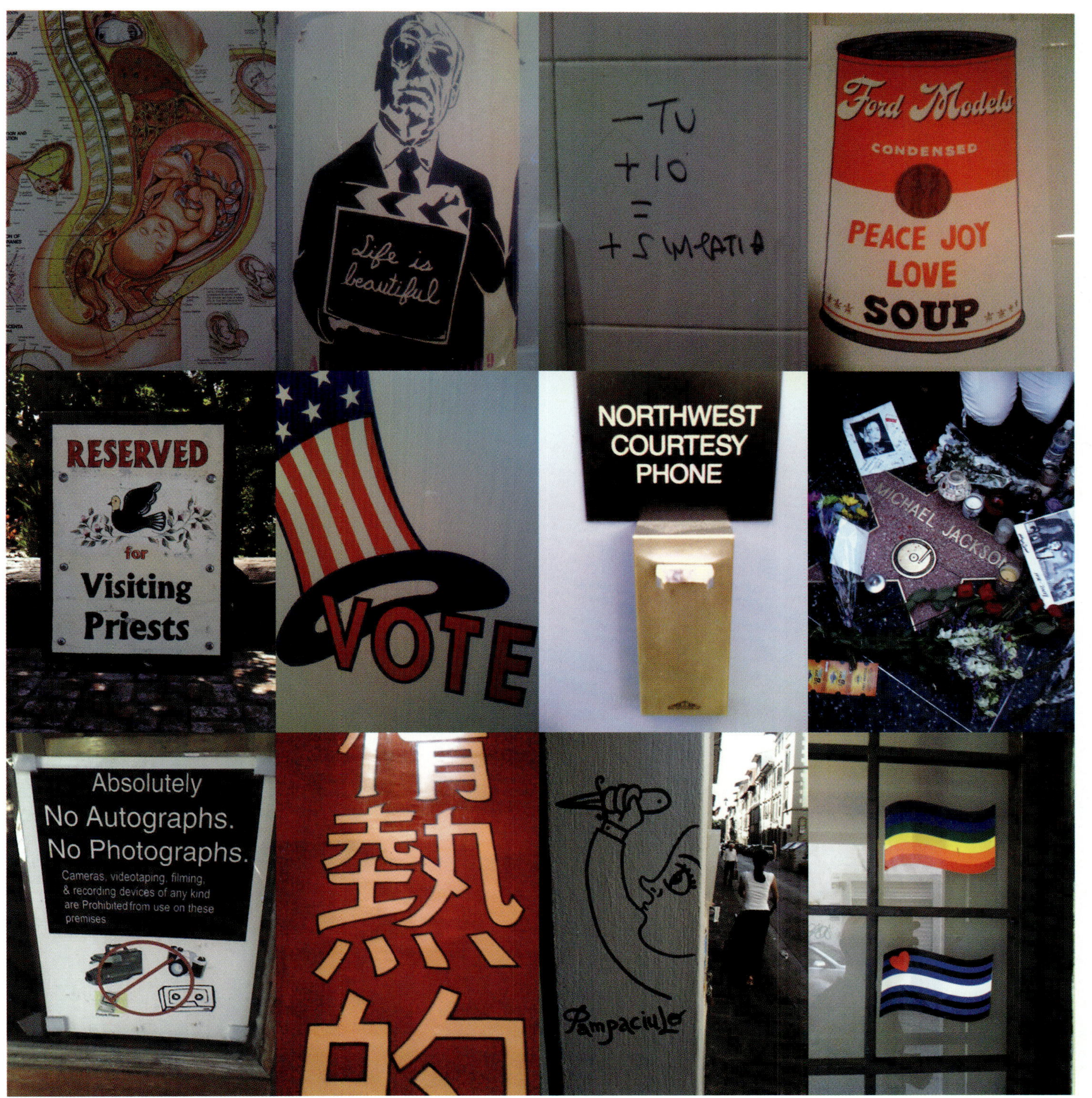
Life is
beautiful
Ford Models
CONDENSED
PEACE JOY
LOVE
SOUP
RESERVED
for
Visiting
Priests
VOTE
NORTHWEST
COURTESY
PHONE
MICHAEL JACKSON
Absolutely
No Autographs.
No Photographs.
Cameras, videotaping, filming,
& recording devices of any kind
are Prohibited from use on these
premises
Pampaciulo

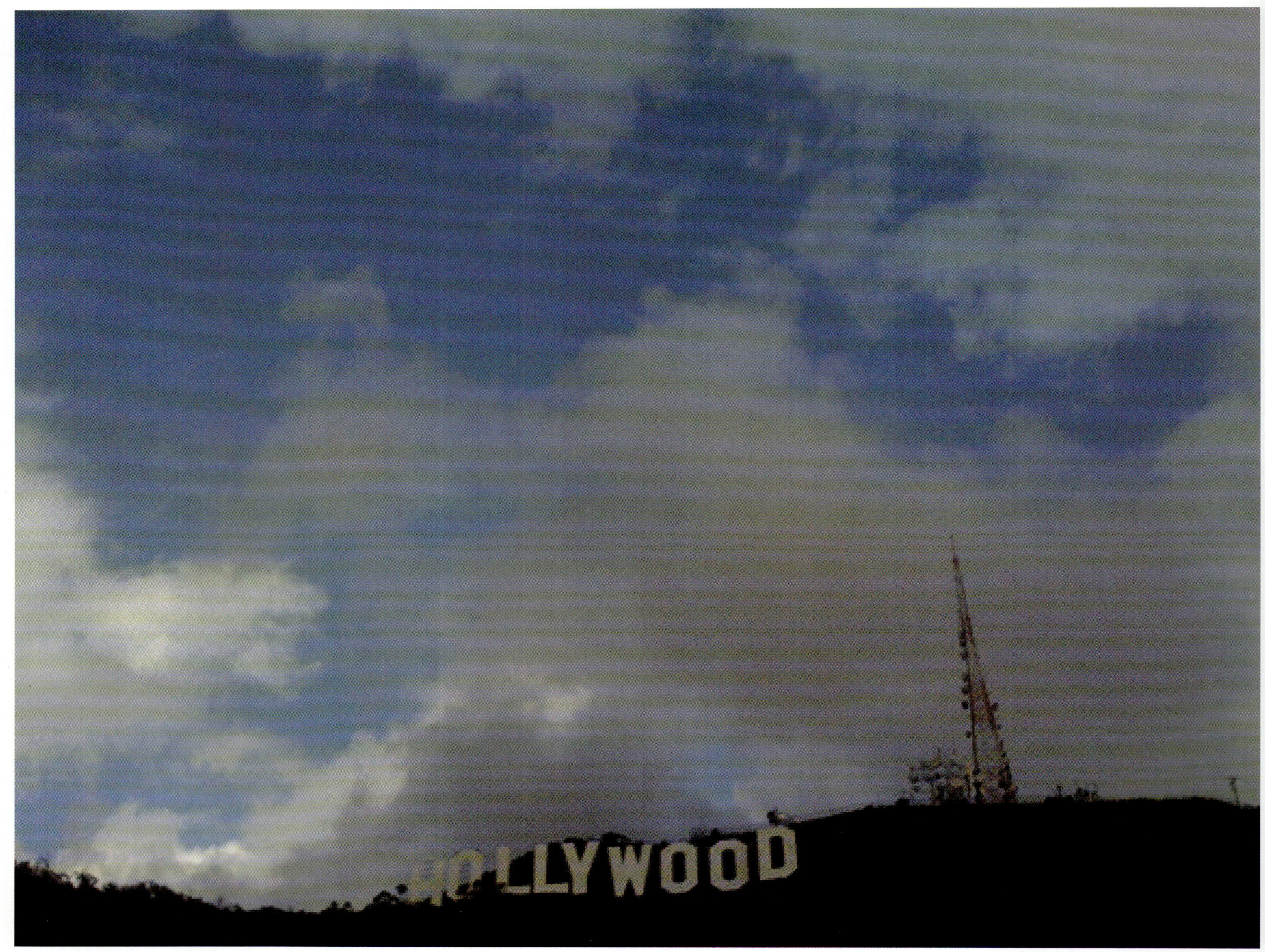

11 Communication

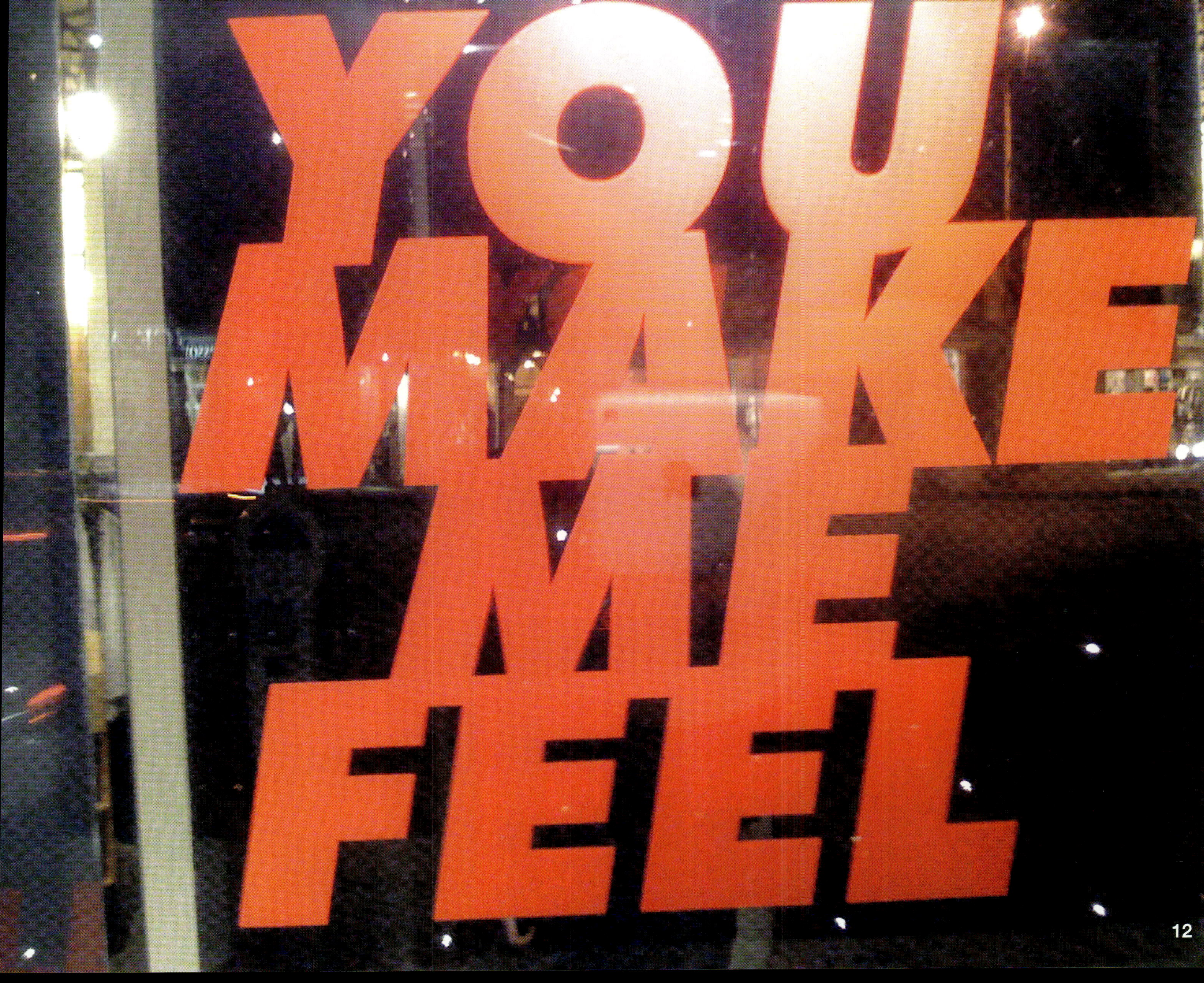

YOU
MAKE
ME
FEEL

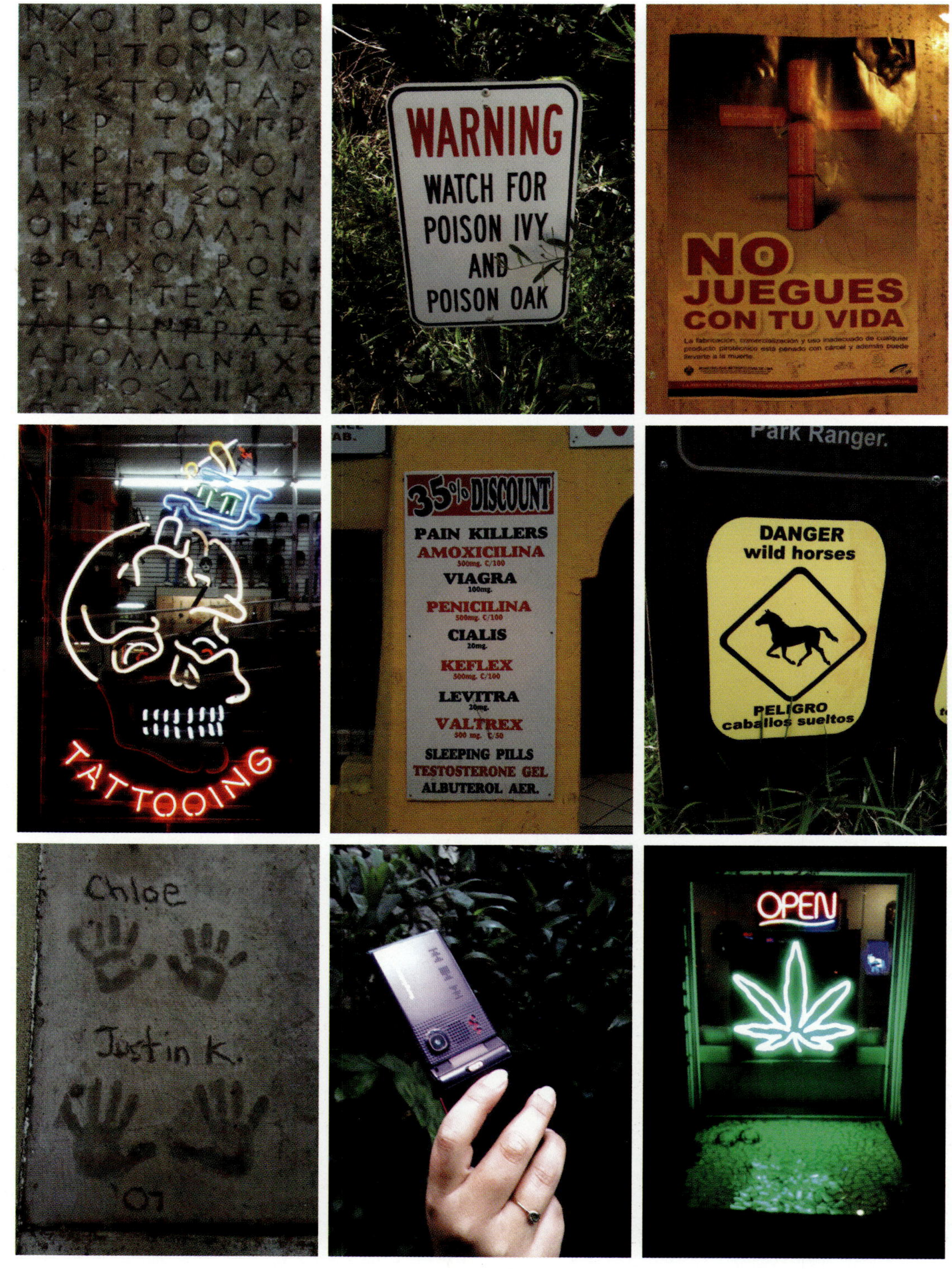

WARNING
WATCH FOR
POISON IVY
AND
POISON OAK
NO
JUEGUES
CON TU VIDA
TATTOOING
35% DISCOUNT
PAIN KILLERS
AMOXICILINA
VIAGRA
PENICILINA
CIALIS
KEFLEX
LEVITRA
VALTREX
SLEEPING PILLS
TESTOSTERONE GEL
ALBUTEROL AER.
Park Ranger.
DANGER
wild horses
PELIGRO
caballos sueltos
Chloe
Justin K.
OPEN

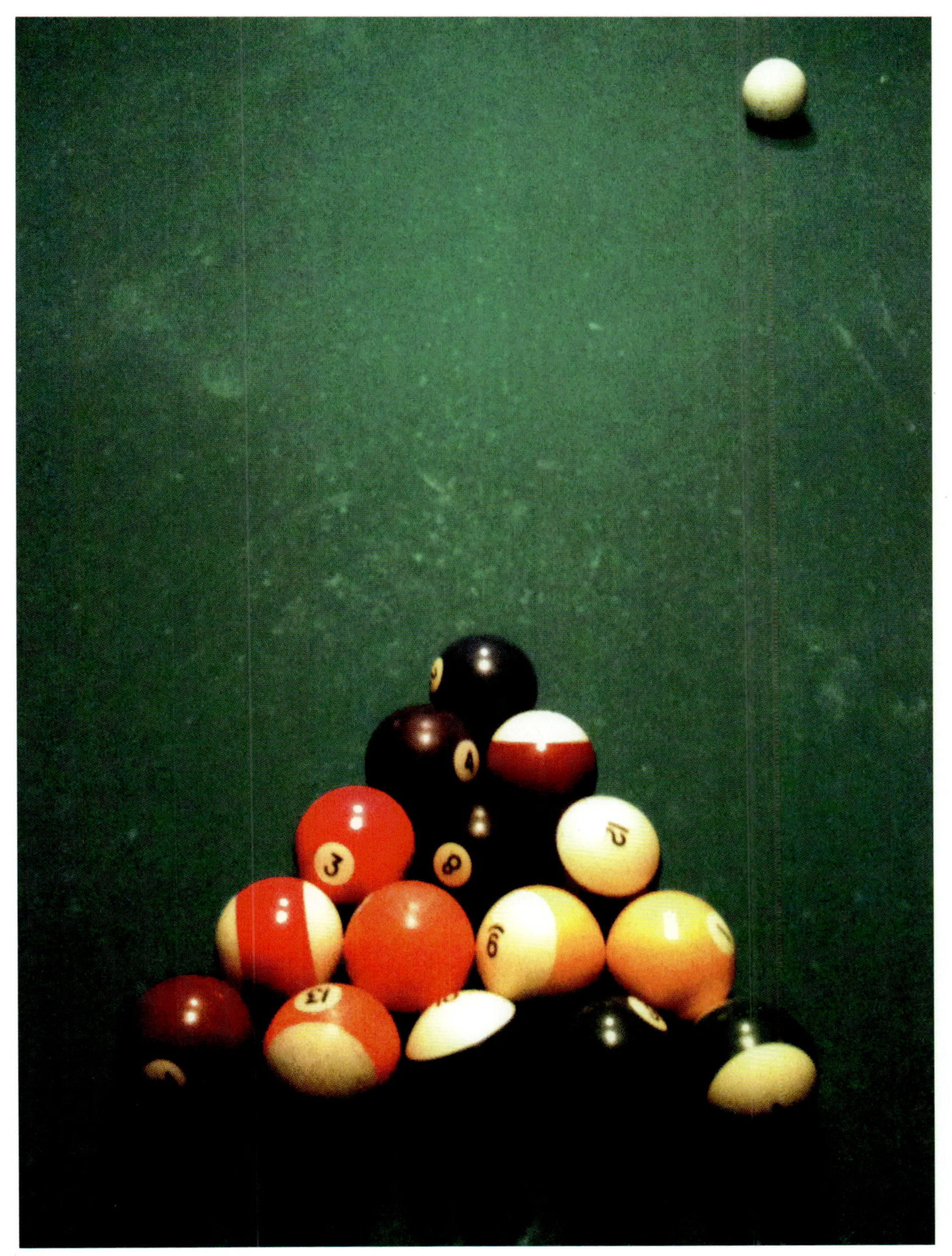

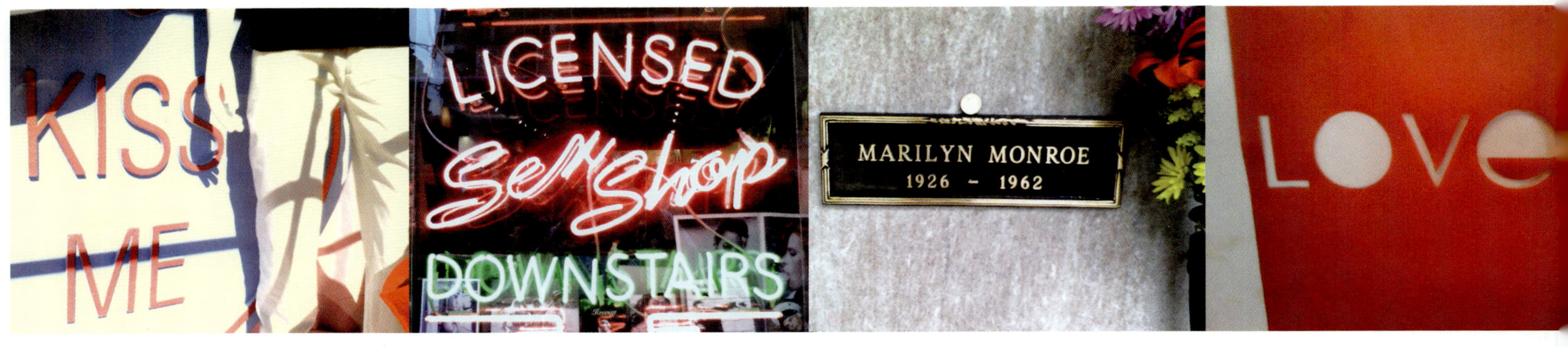

KISS ME
LICENSED Sex Shop DOWNSTAIRS
MARILYN MONROE
1926 - 1962
LOVE

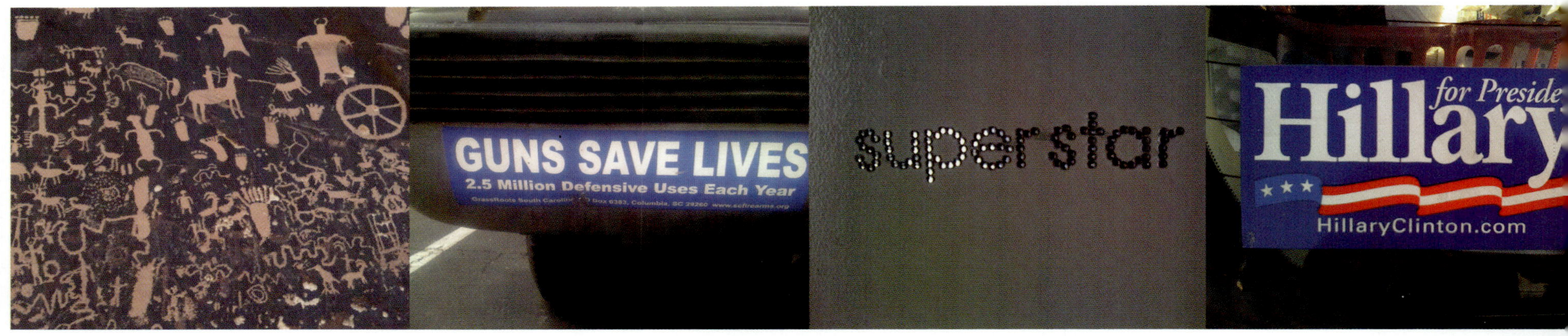

GUNS SAVE LIVES
2.5 Million Defensive Uses Each Year
superstar
Hillary for Preside
HillaryClinton.com

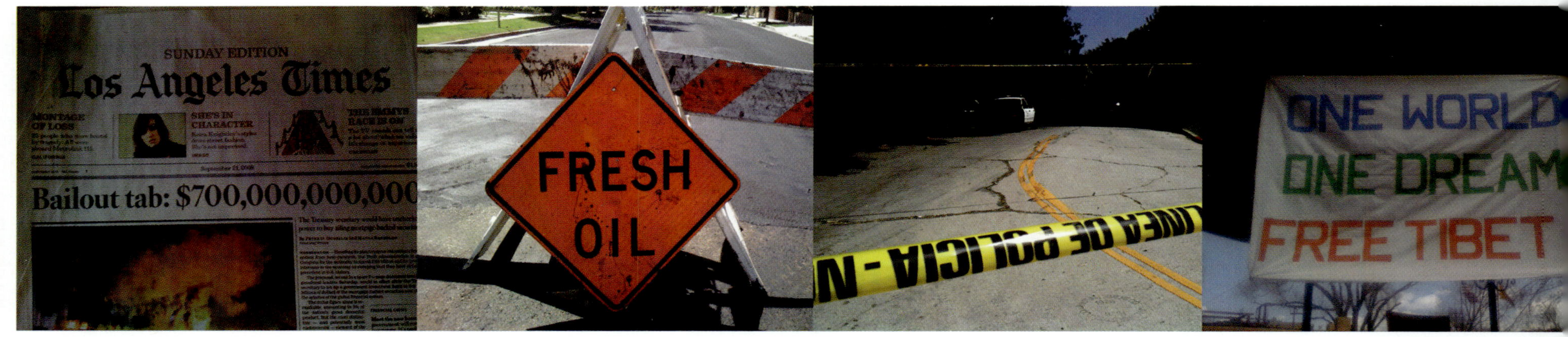

SUNDAY EDITION
Los Angeles Times
SHE'S IN CHARACTER
Bailout tab: $700,000,000,000
FRESH OIL
LINEA DE POLICIA
ONE WORLD
ONE DREAM
FREE TIBET

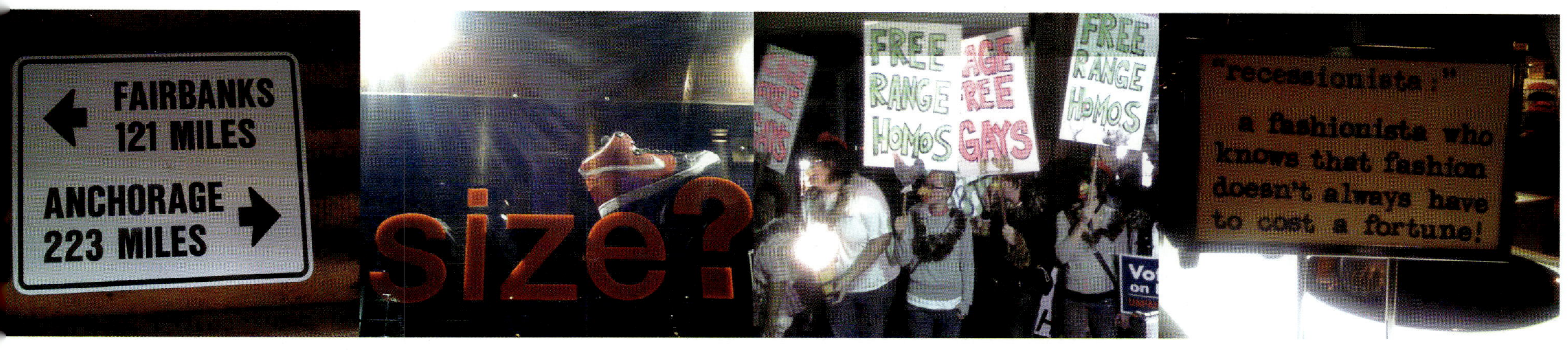

FAIRBANKS
121 MILES
ANCHORAGE
223 MILES
size?
FREE
RANGE
HOMOS
AGE
REE
GAYS
FREE
RANGE
HOMOS
Vote
on
"recessionista:"
a fashionista who
knows that fashion
doesn't always have
to cost a fortune!

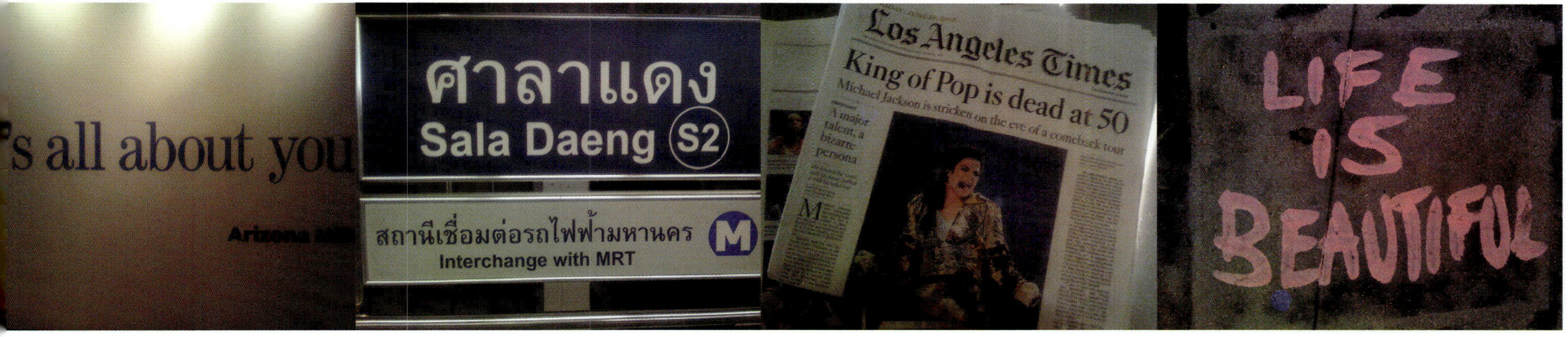

's all about you
Arizona
ศาลาแดง
Sala Daeng S2
สถานีเชื่อมต่อรถไฟฟ้ามหานคร M
Interchange with MRT
Los Angeles Times
King of Pop is dead at 50
Michael Jackson is stricken on the eve of a comeback tour
A major
talent, a
bizarre
persona
LIFE
IS
BEAUTIFUL

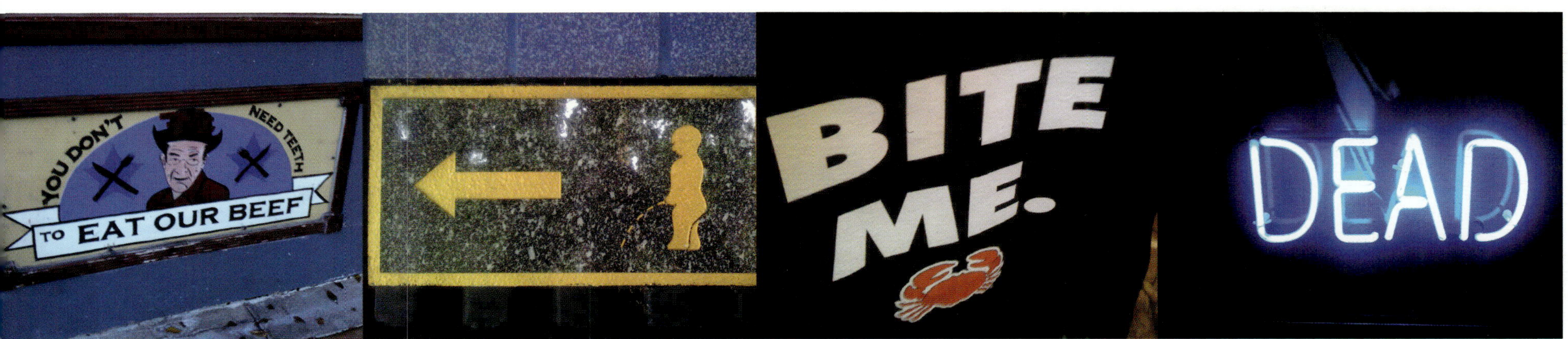

YOU DON'T
NEED TEETH
TO EAT OUR BEEF
BITE
ME.
DEAD

measuring
your

own grave

THEY
ARE
WATCHING
YOU

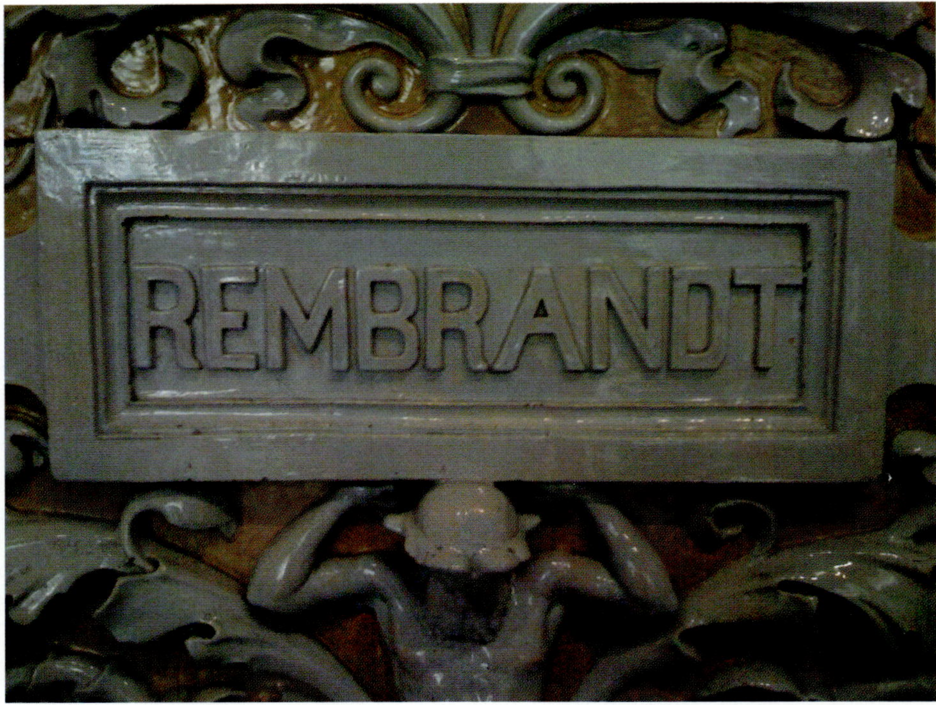
REMBRANDT

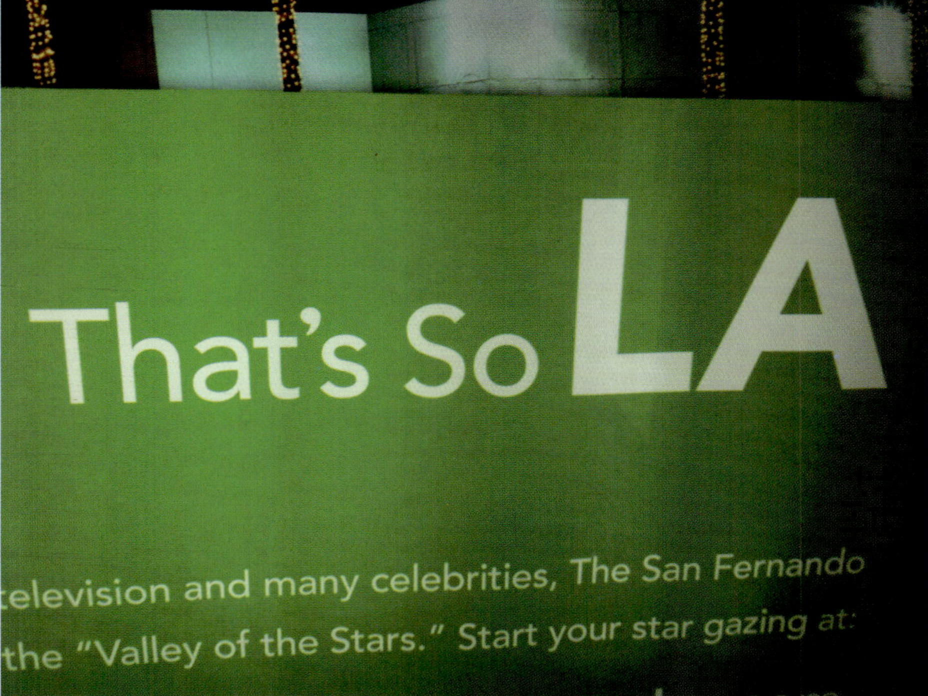
That's So LA
television and many celebrities, The San Fernando
the "Valley of the Stars." Start your star gazing at:

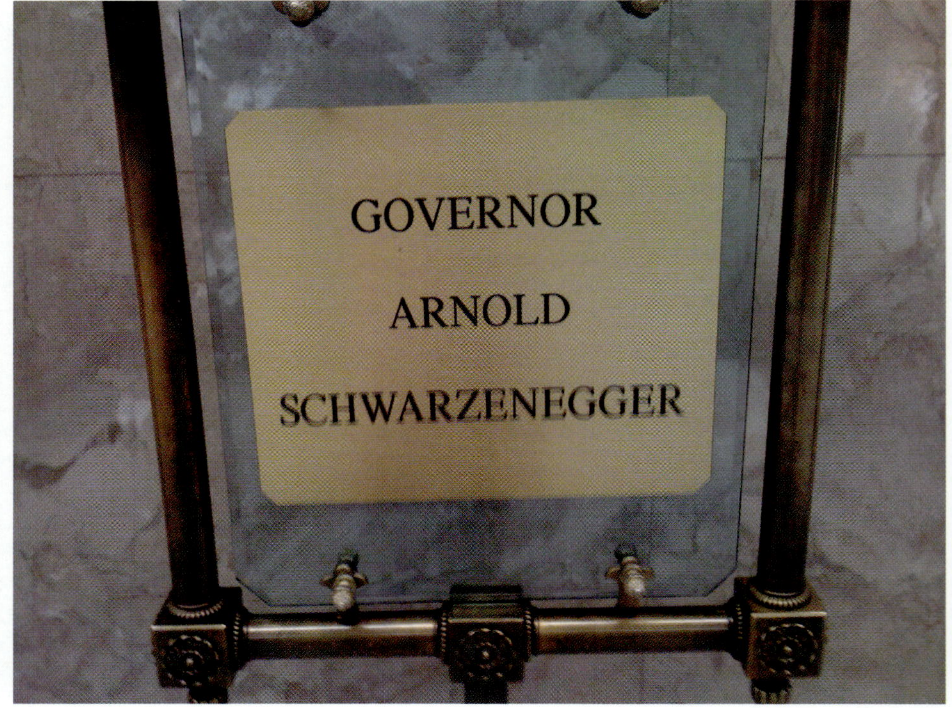
GOVERNOR

ARNOLD

SCHWARZENEGGER

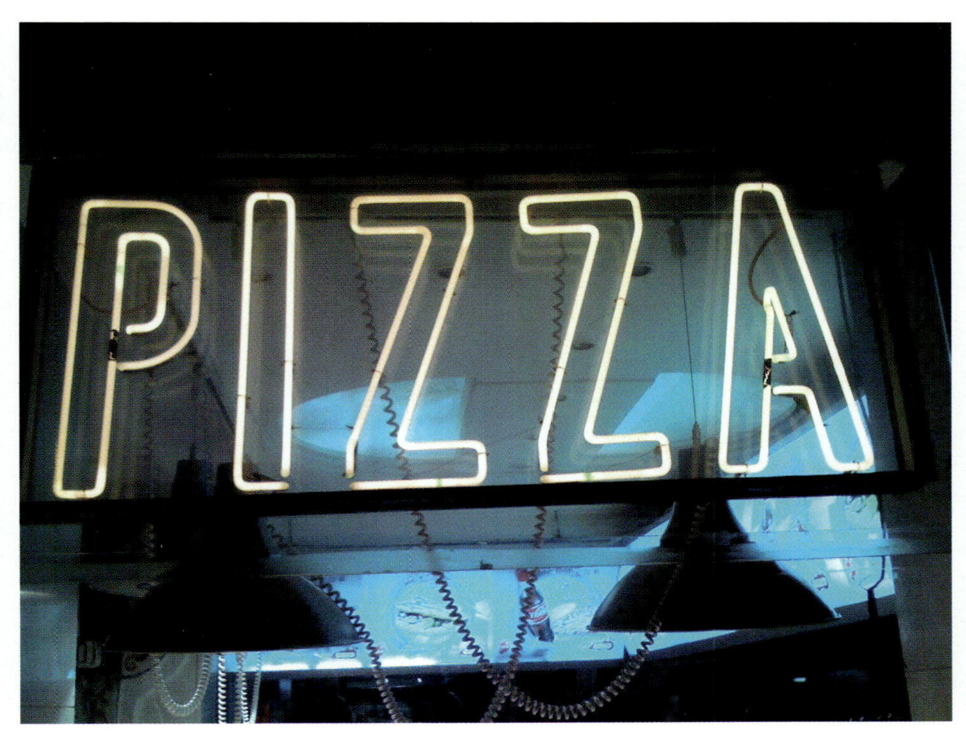
PIZZA

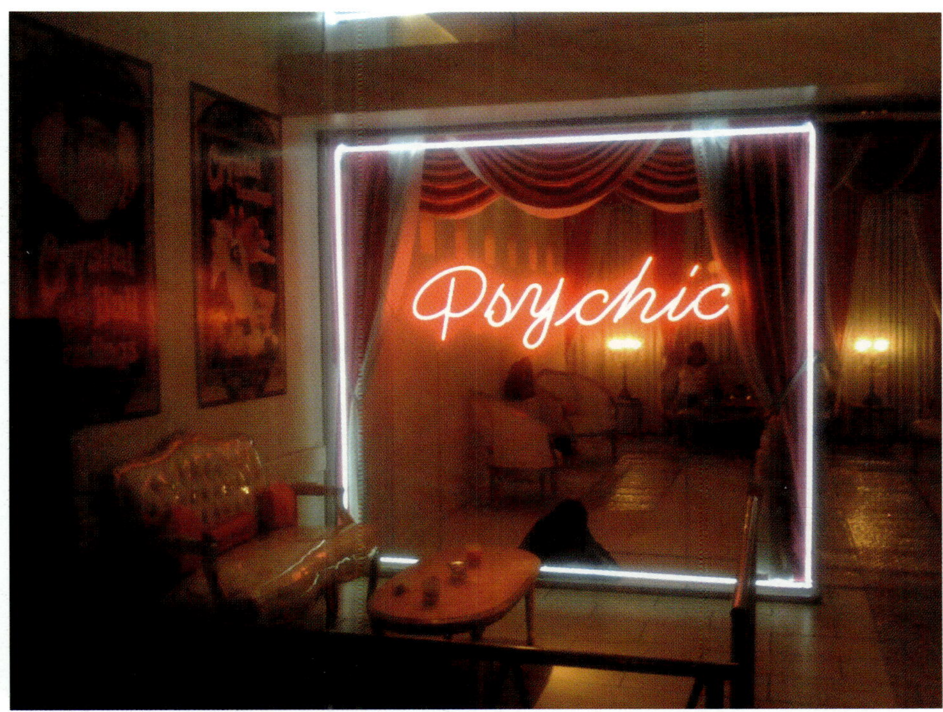
Psychic

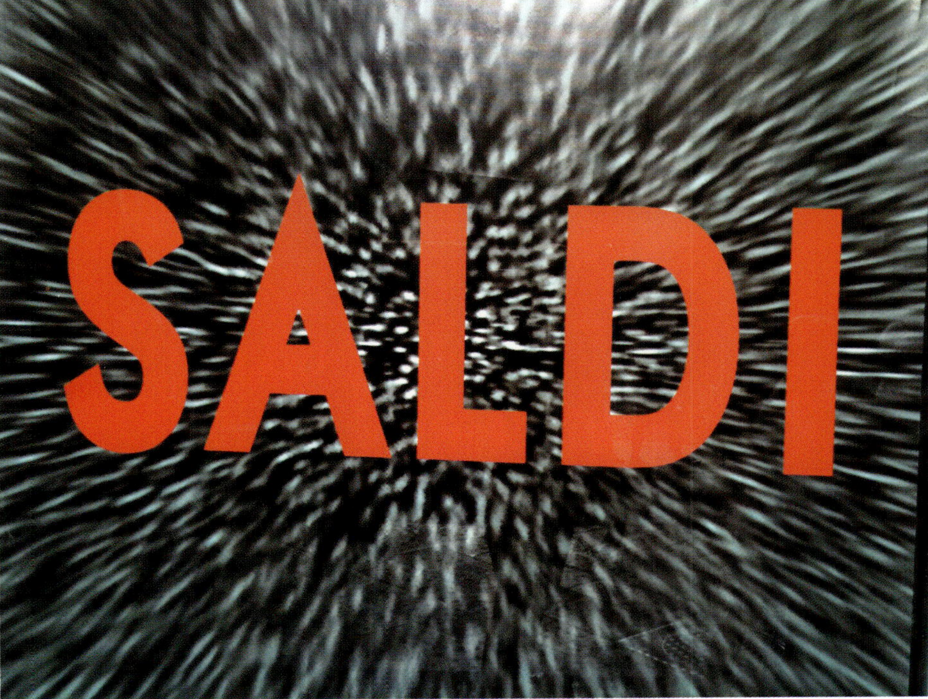
SALDI

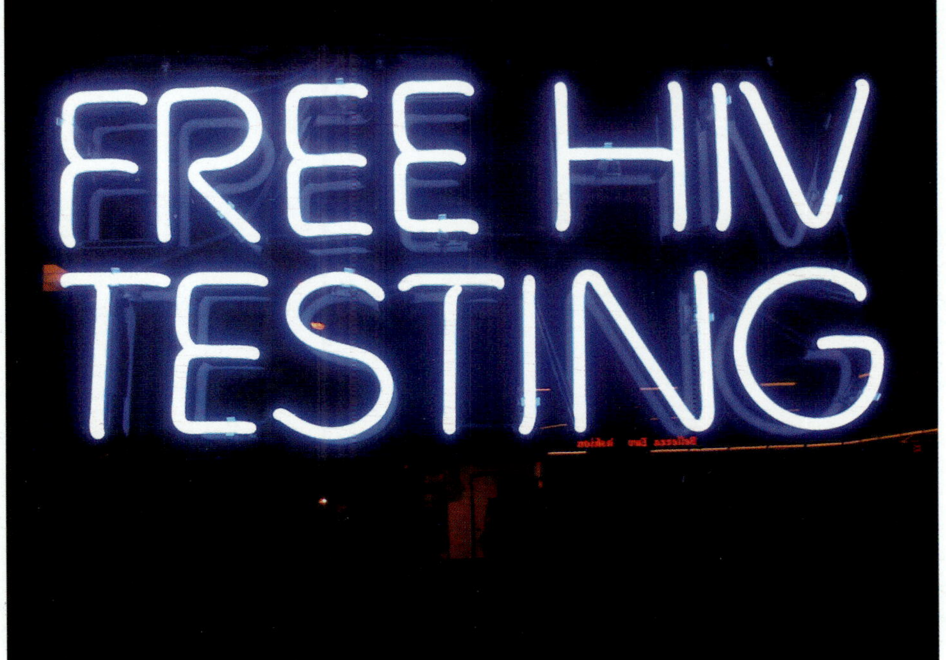
FREE HIV
TESTING

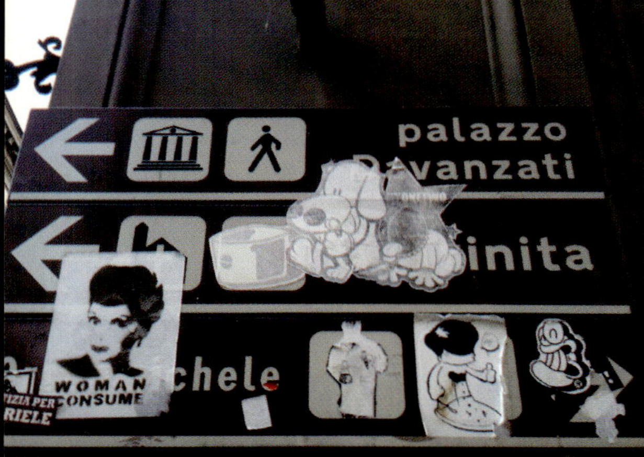

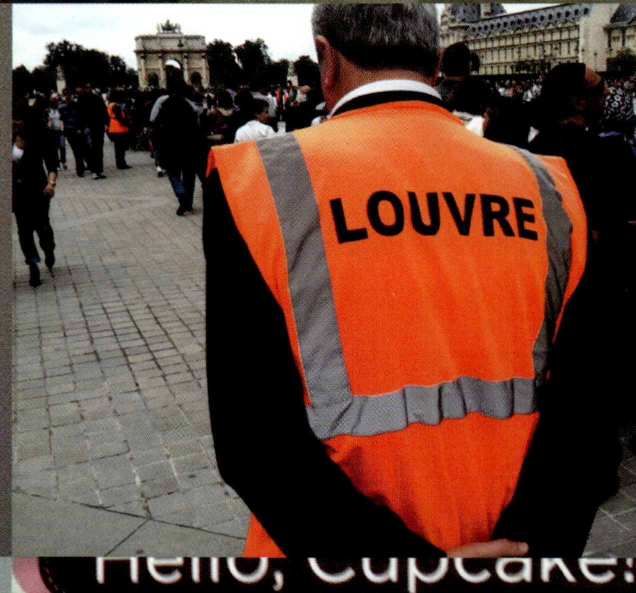

GRAND HYATT CAIRO
Evacuation Assembly
Point No 1
نقطة تجمع وإخلاء رقم : ١

ستر ة النجاة تحت المقعد
اربط حزام المقعد أثناء جلوسك

Life vest under your seat
Fasten seat belt while seated

NO WAR

POST NO BILLS

Hello, Cupcake!
Hello, Cupcake!
Hello, Cupcake!
Hello, Cupcake!
Hello, Cupcake!
Hello, Cupcake!
Hello, Cupcake!

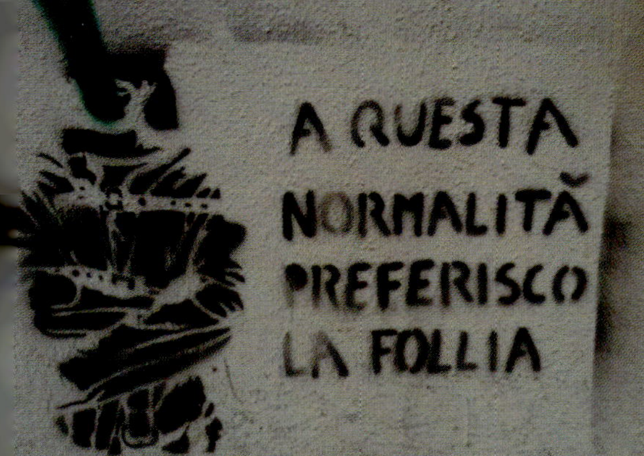

A QUESTA
NORMALITÃ
PREFERISCO
LA FOLLIA

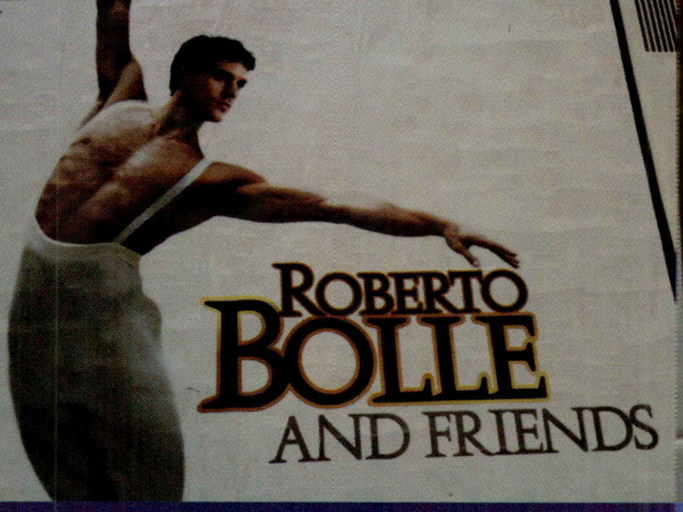

ROBERTO
BOLLE
AND FRIENDS

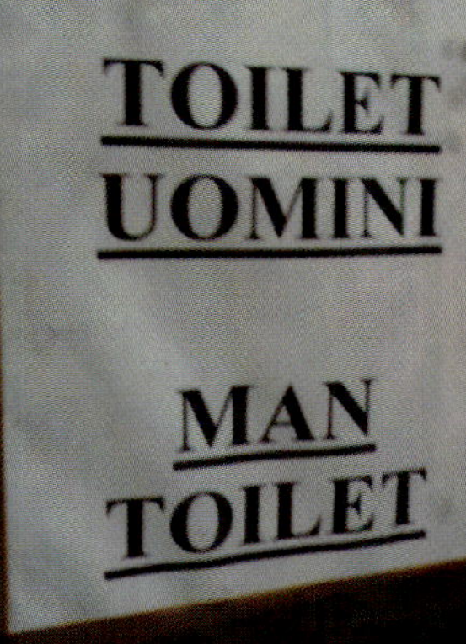

TOILET
UOMINI

MAN
TOILET

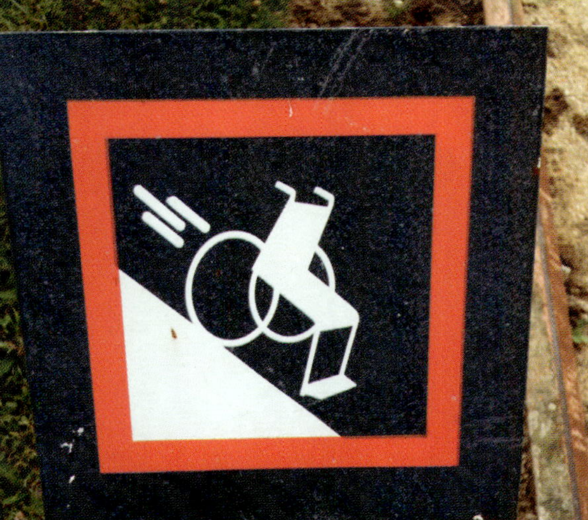

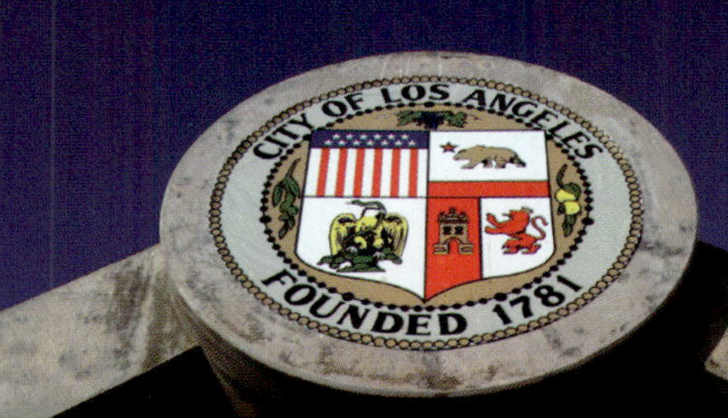

CITY OF LOS ANGELES
FOUNDED 1781

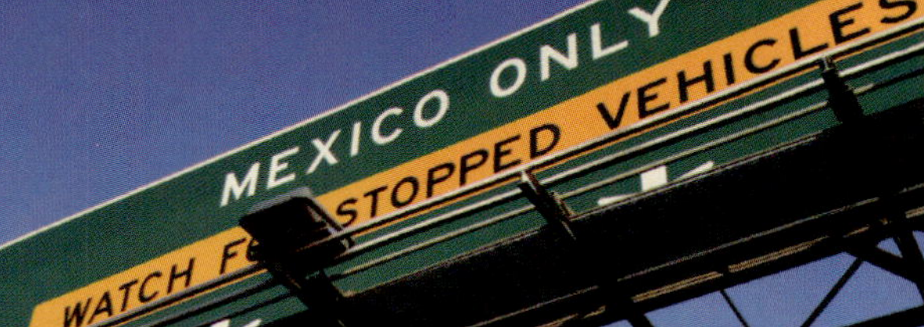

MEXICO ONLY
WATCH FOR STOPPED VEHICLES

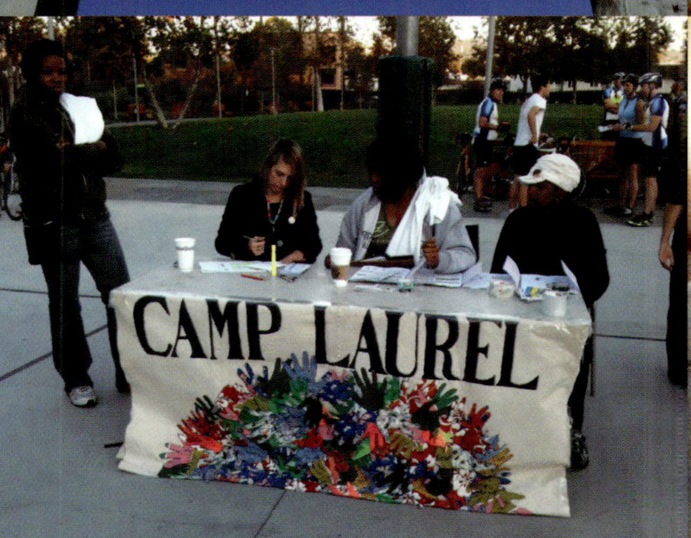

Snake Pit
Ale House
CAMP LAUREL

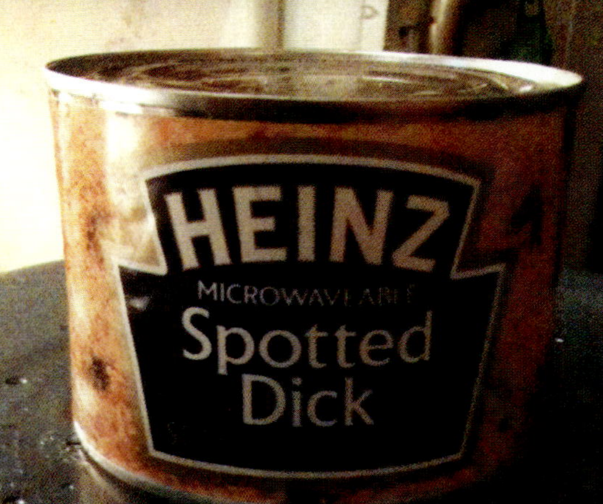

HEINZ
MICROWAVEABLE
Spotted
Dick

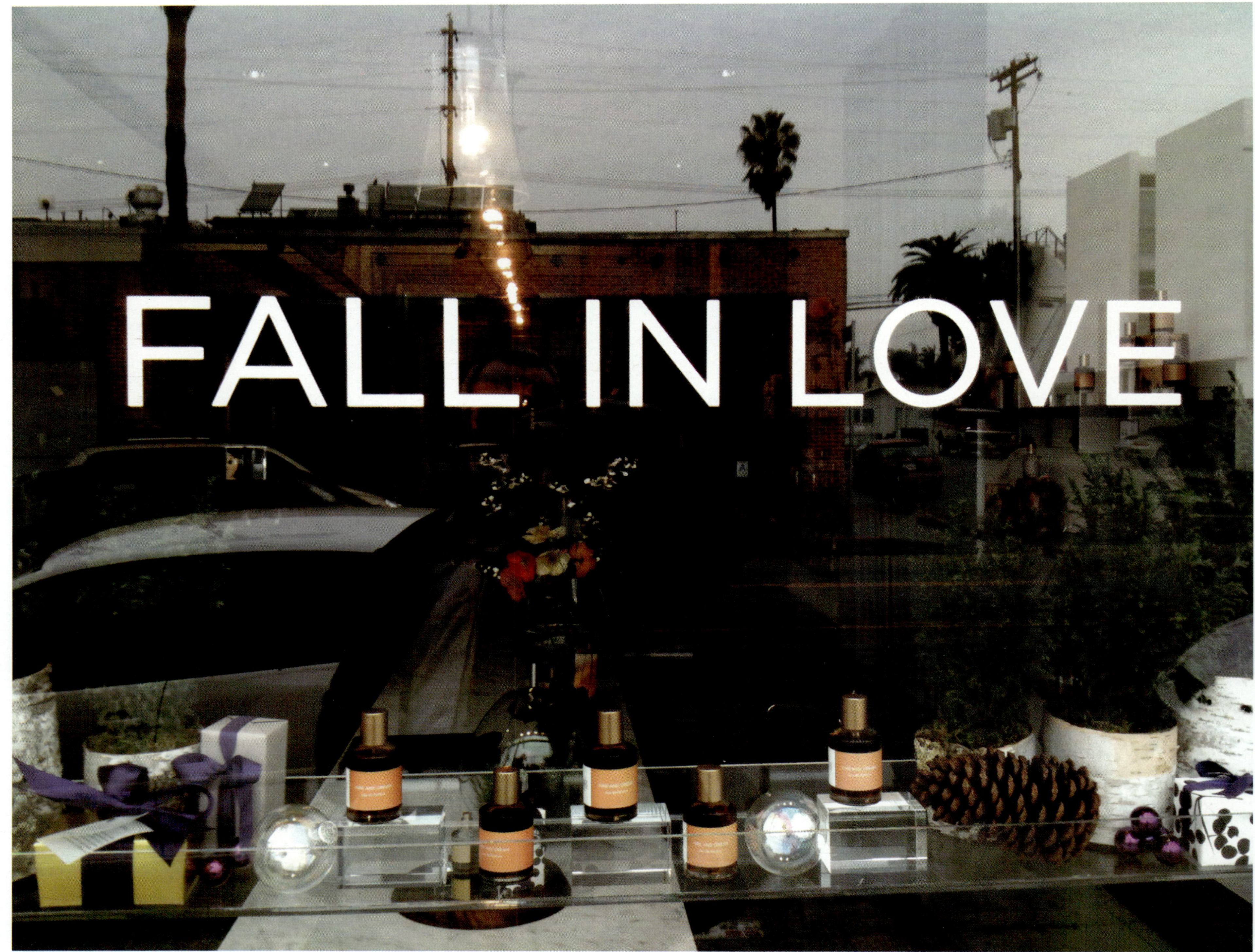

FALL IN LOVE

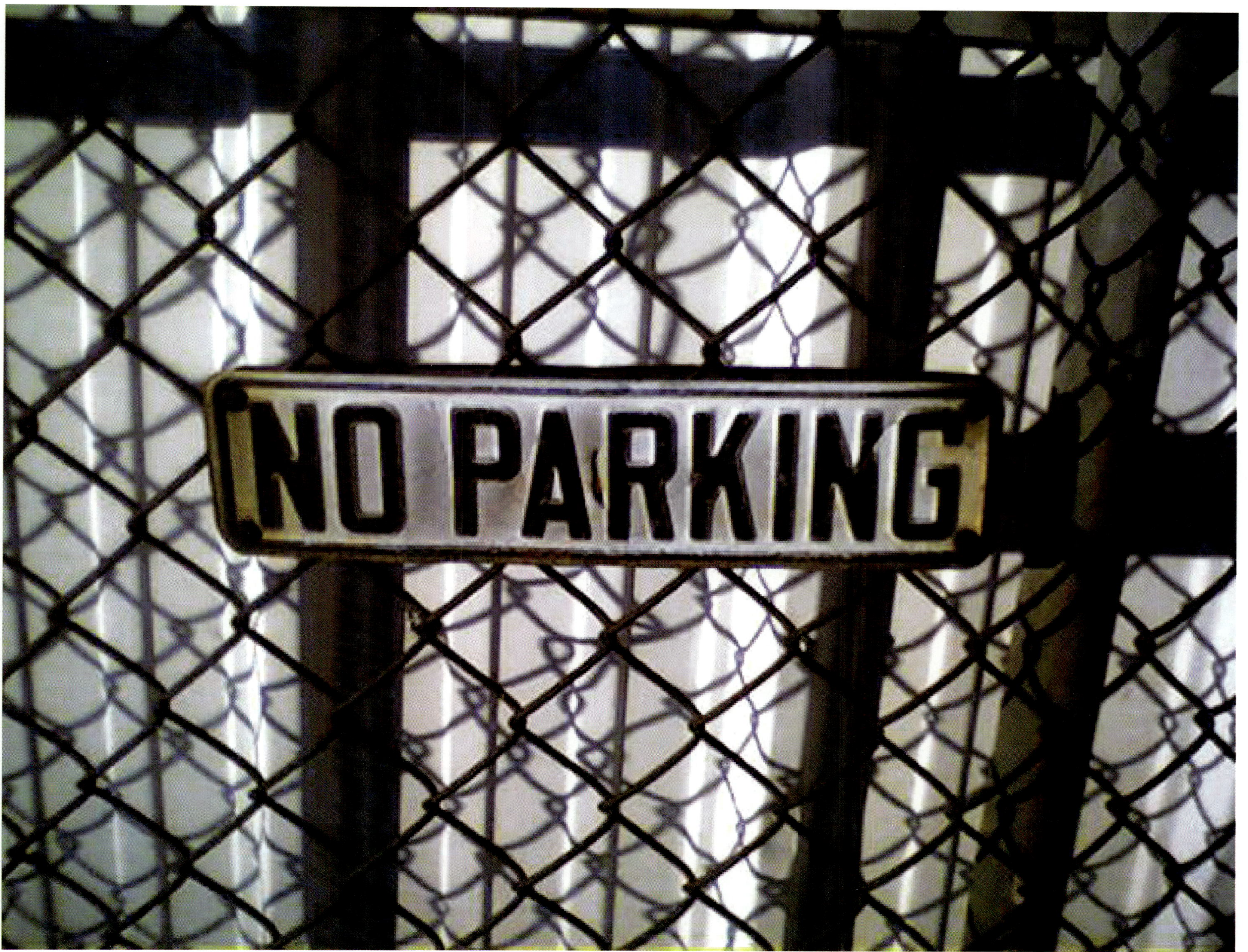
NO PARKING

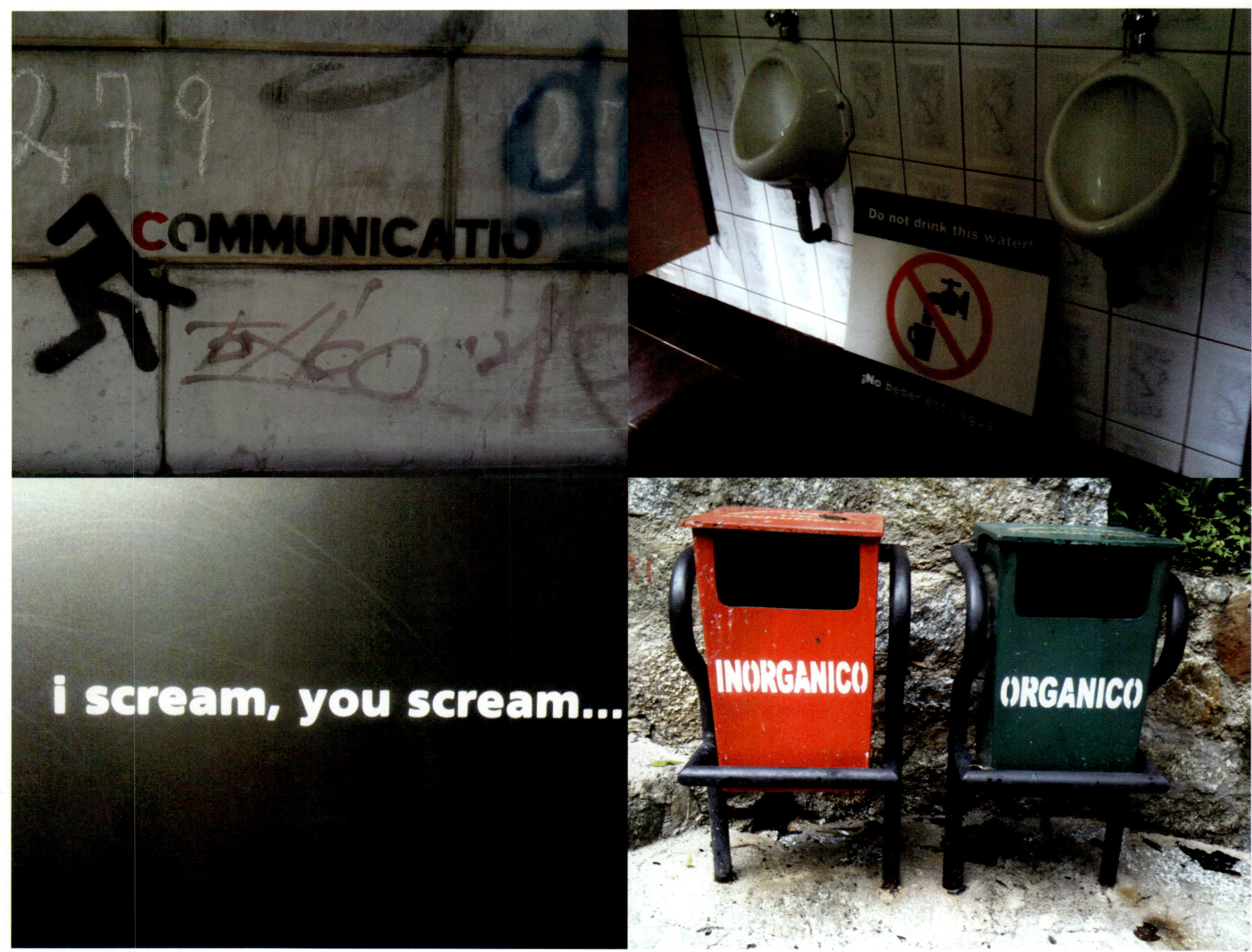
COMMUNICATIO
Do not drink this water
i scream, you scream...
INORGANICO
ORGANICO

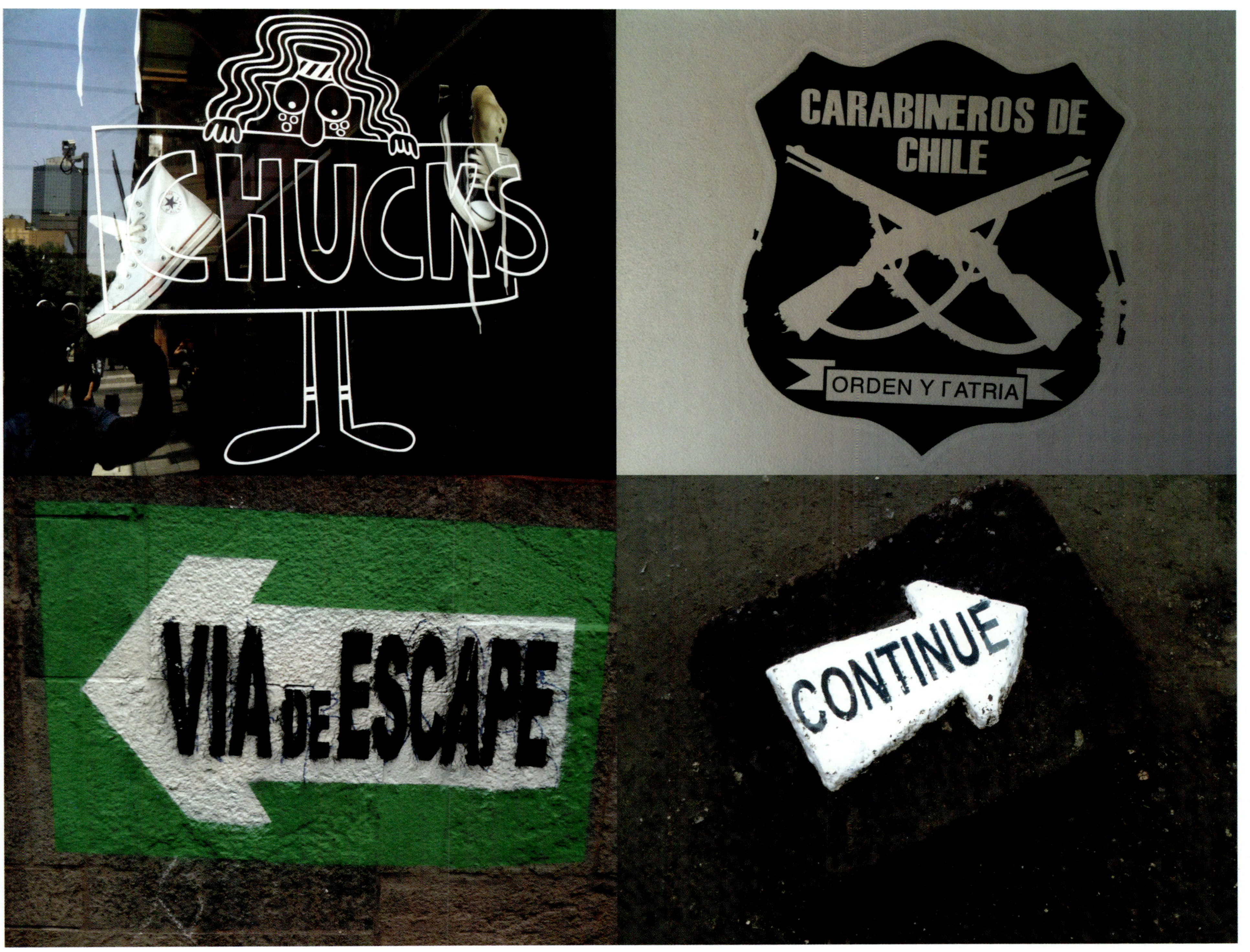
CHUCKS
CARABINEROS DE CHILE
ORDEN Y PATRIA
VIA DE ESCAPE
CONTINUE

FACES

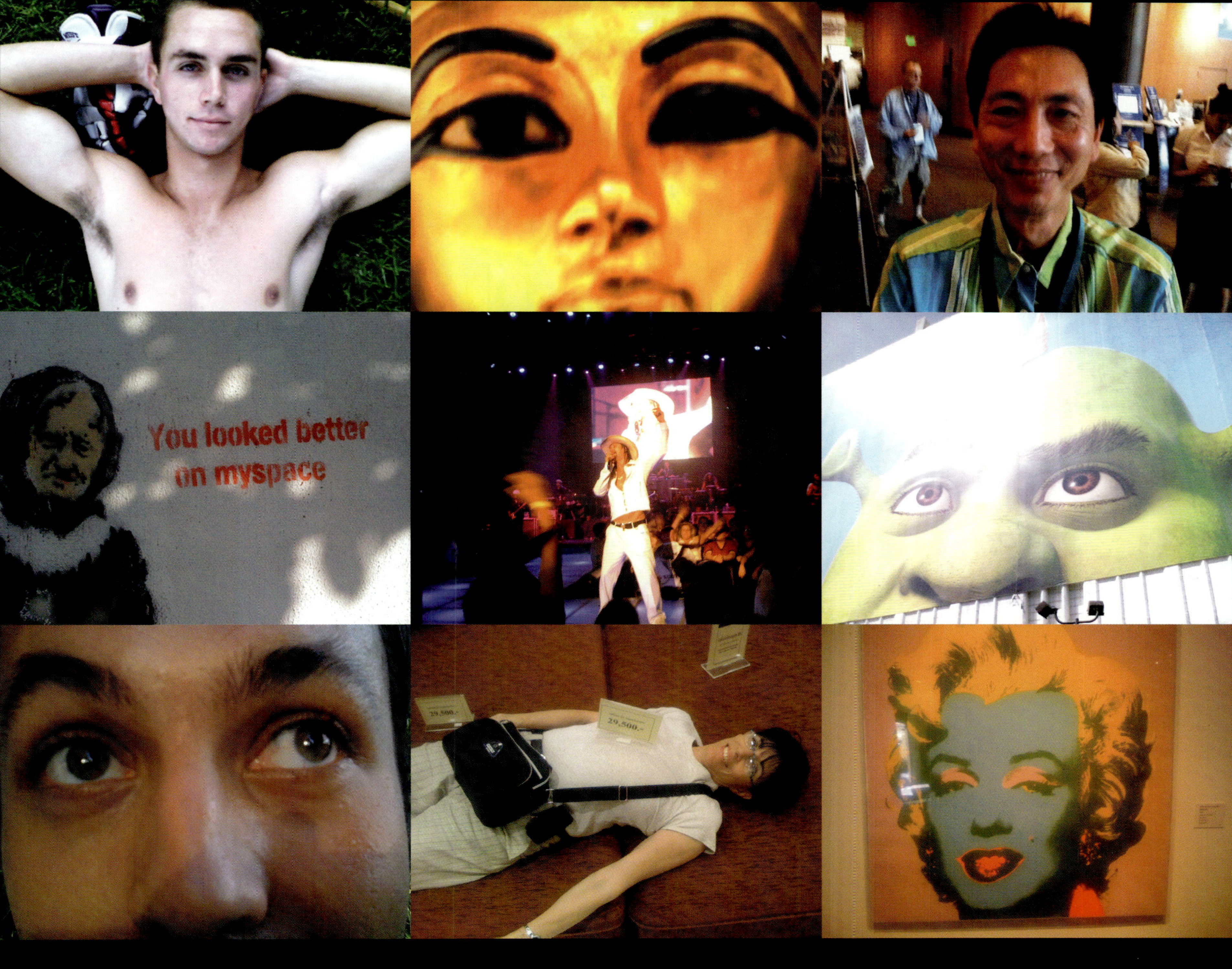
You looked better
on myspace
29,500.-

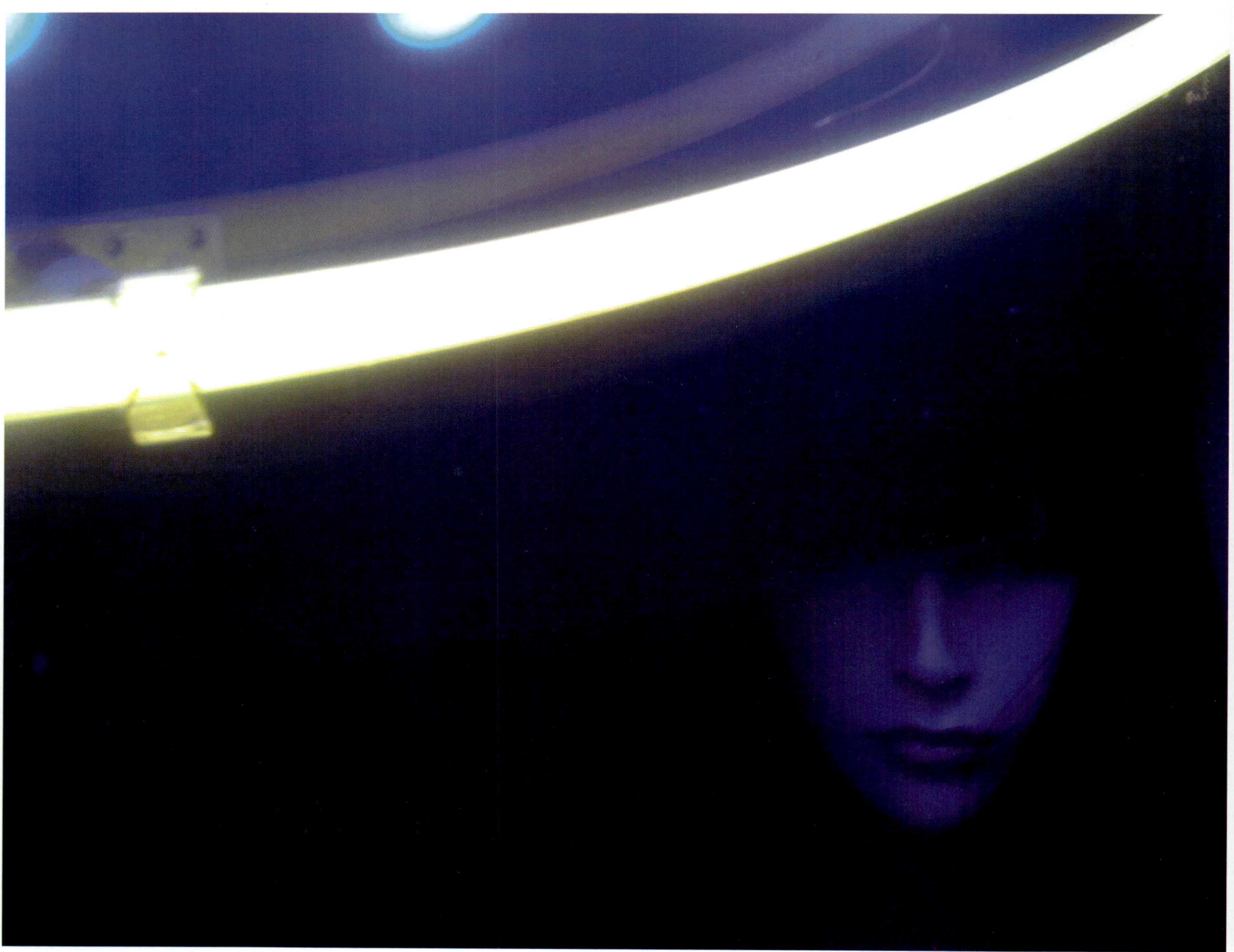

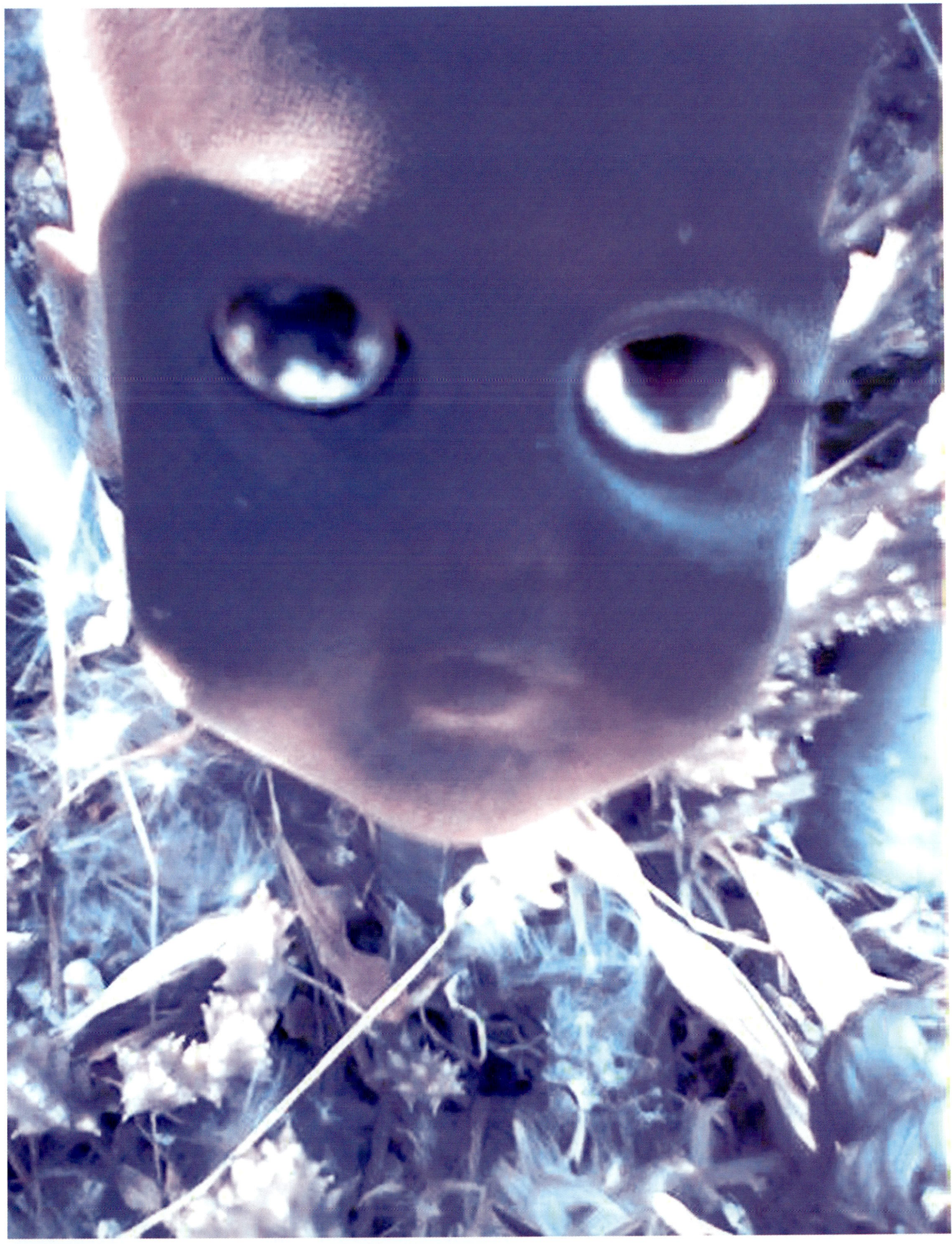

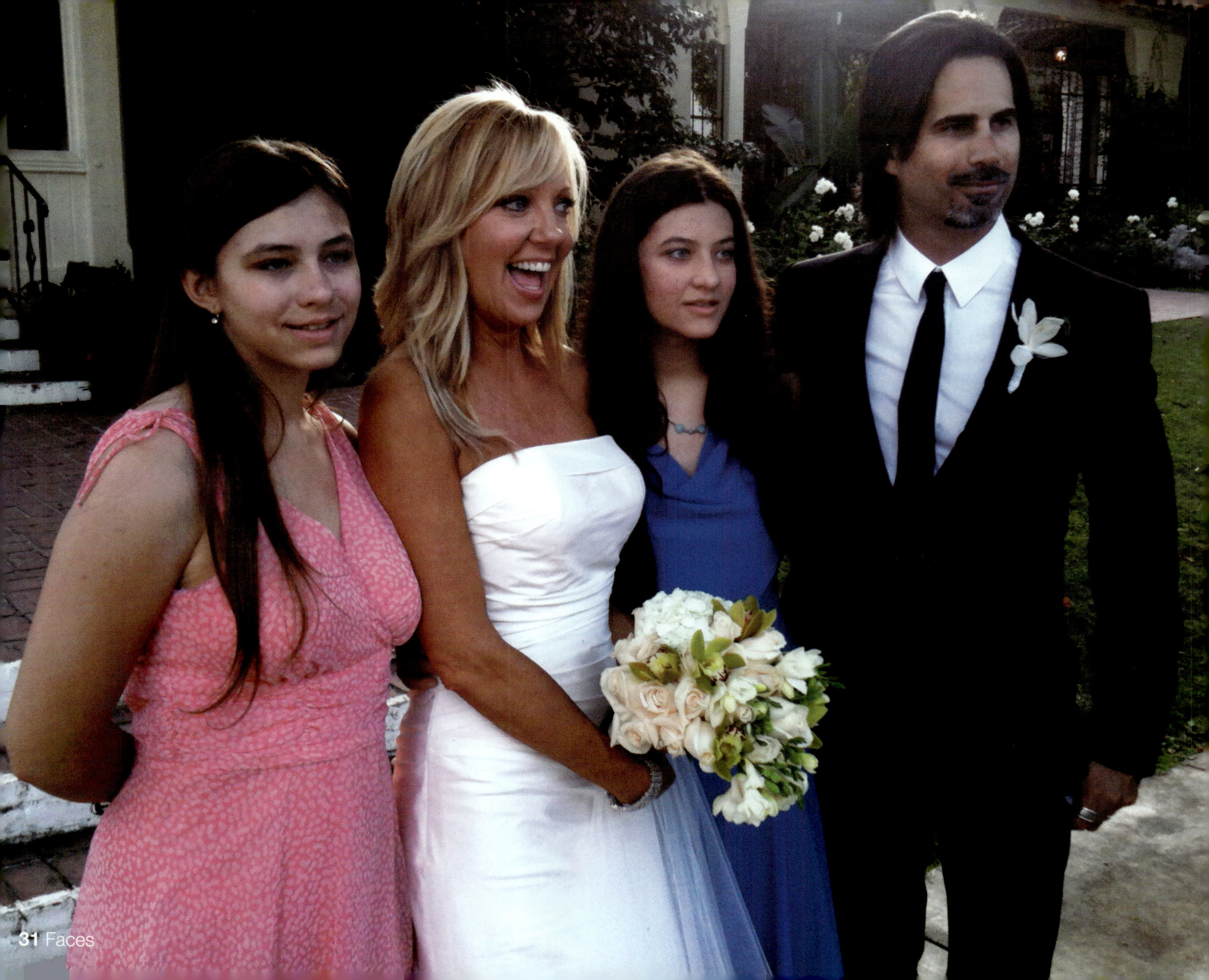

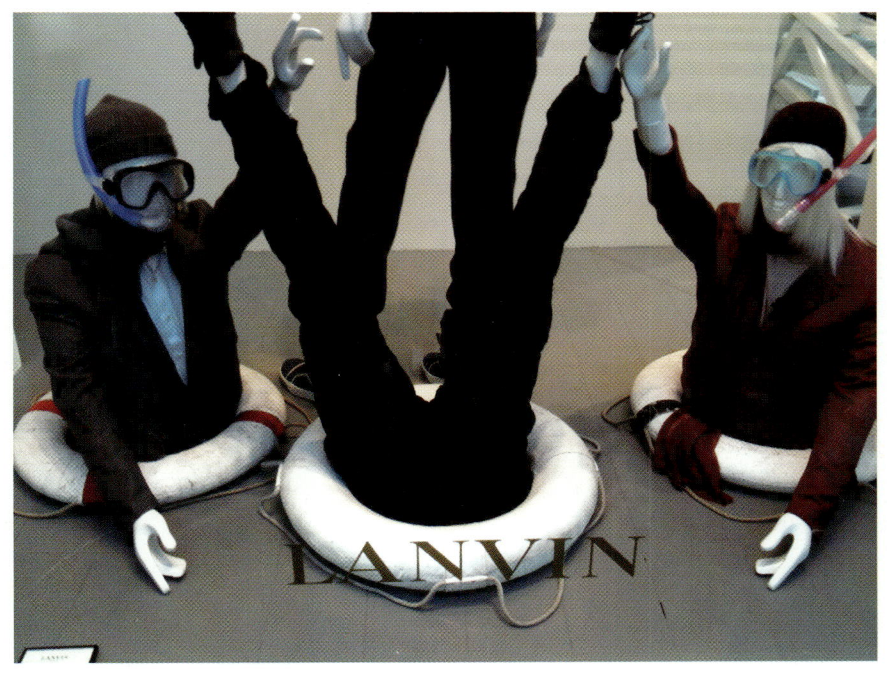

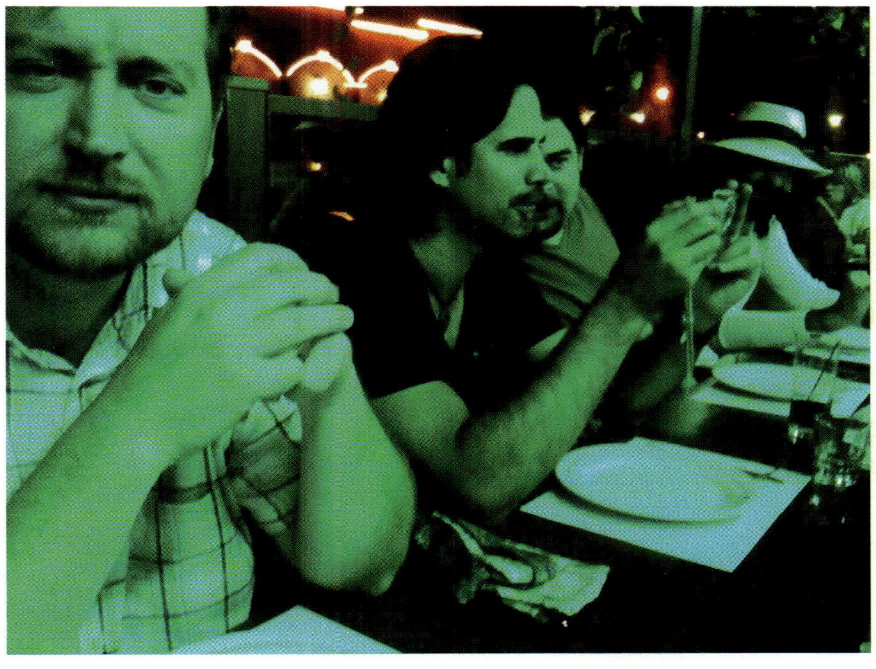

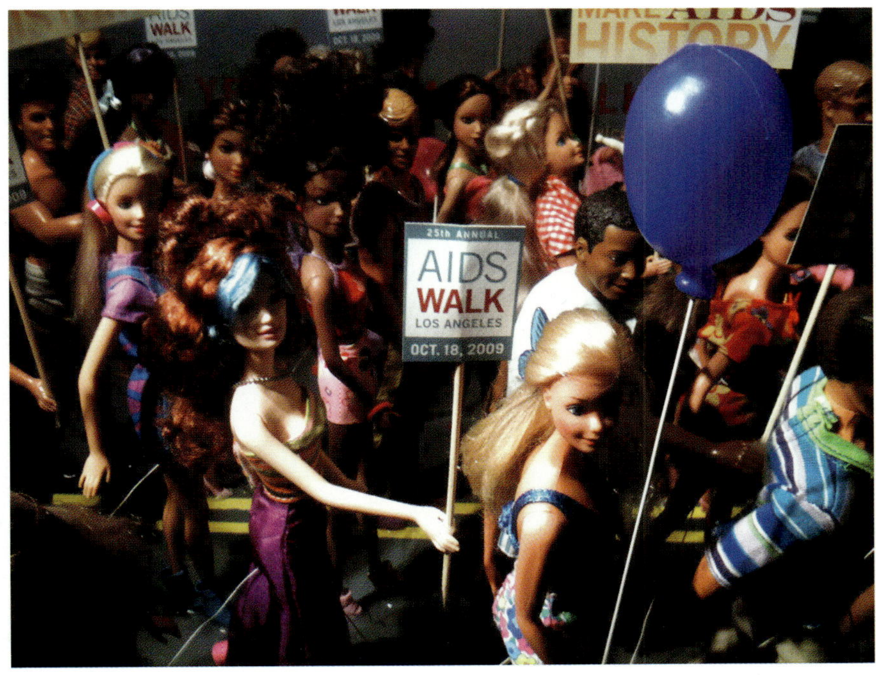

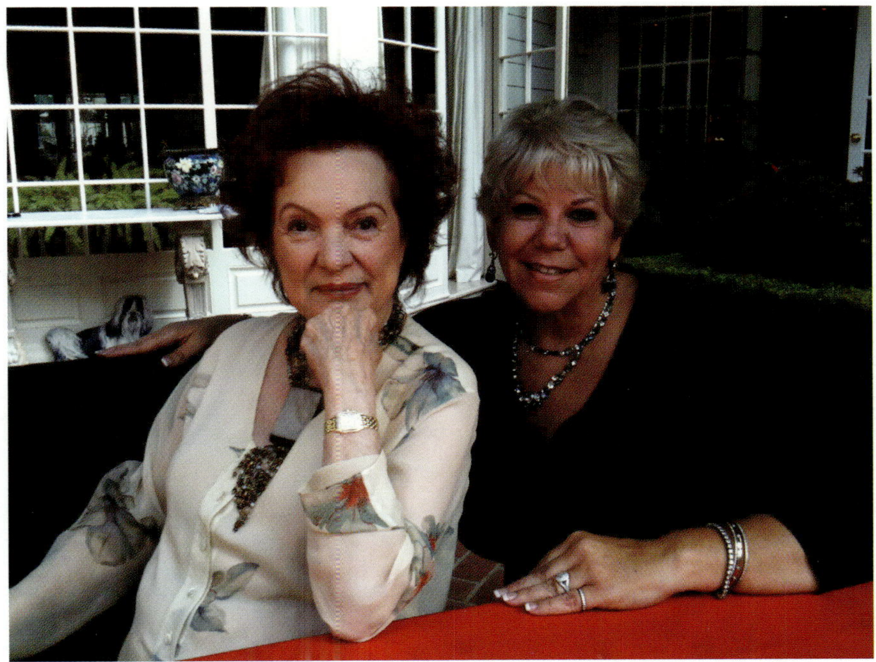

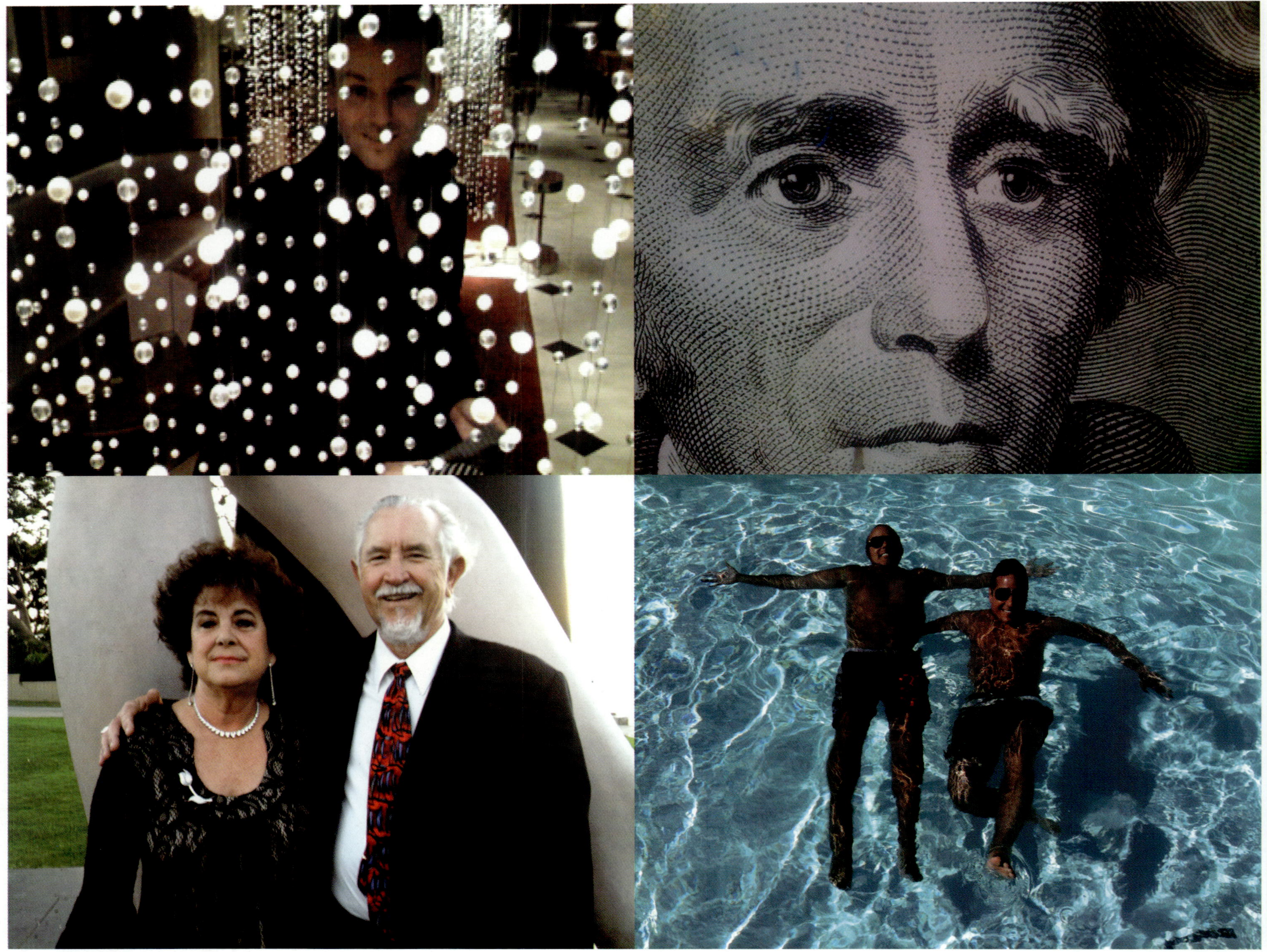

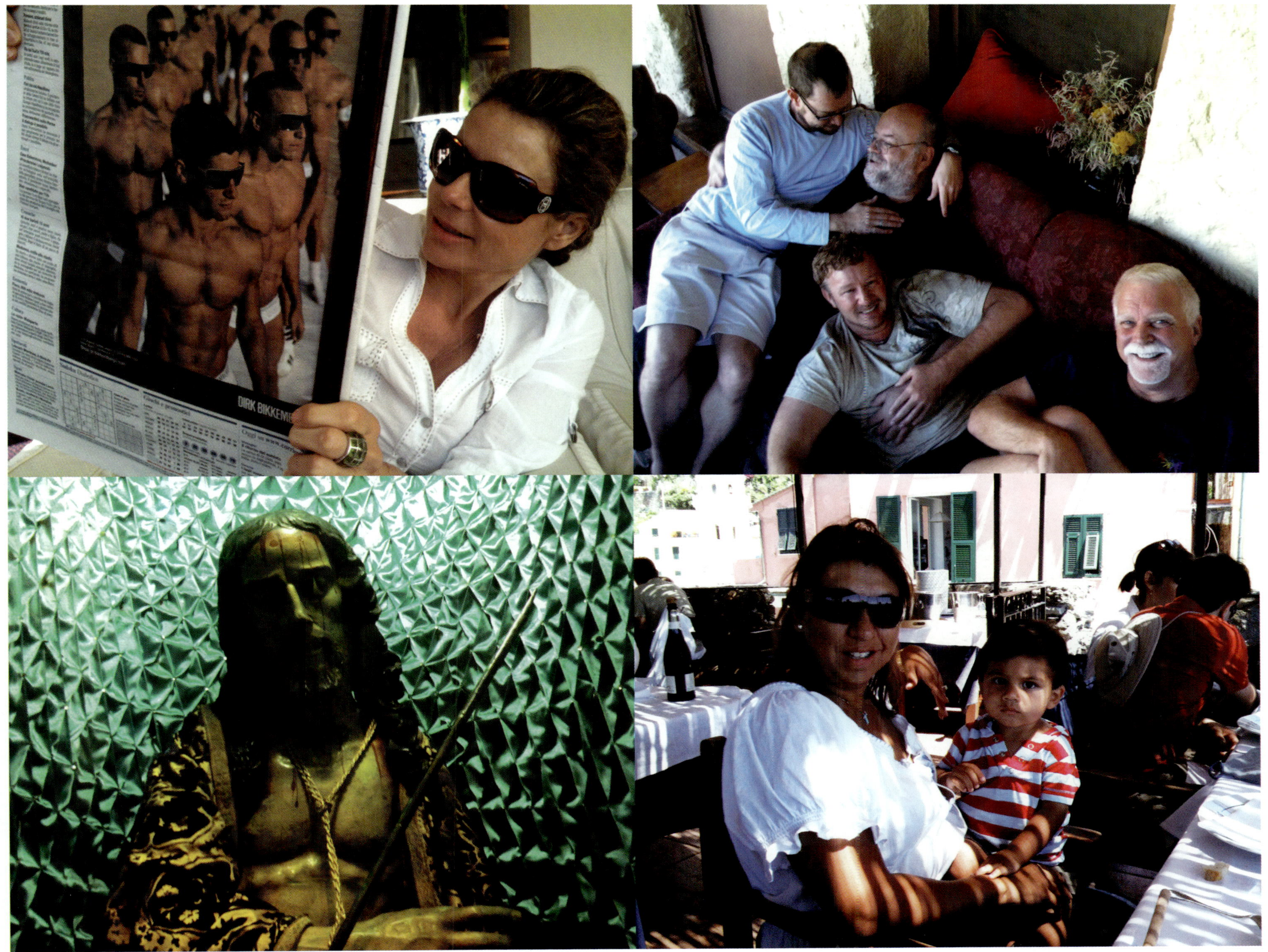

DIRK BIKKEMBERGS

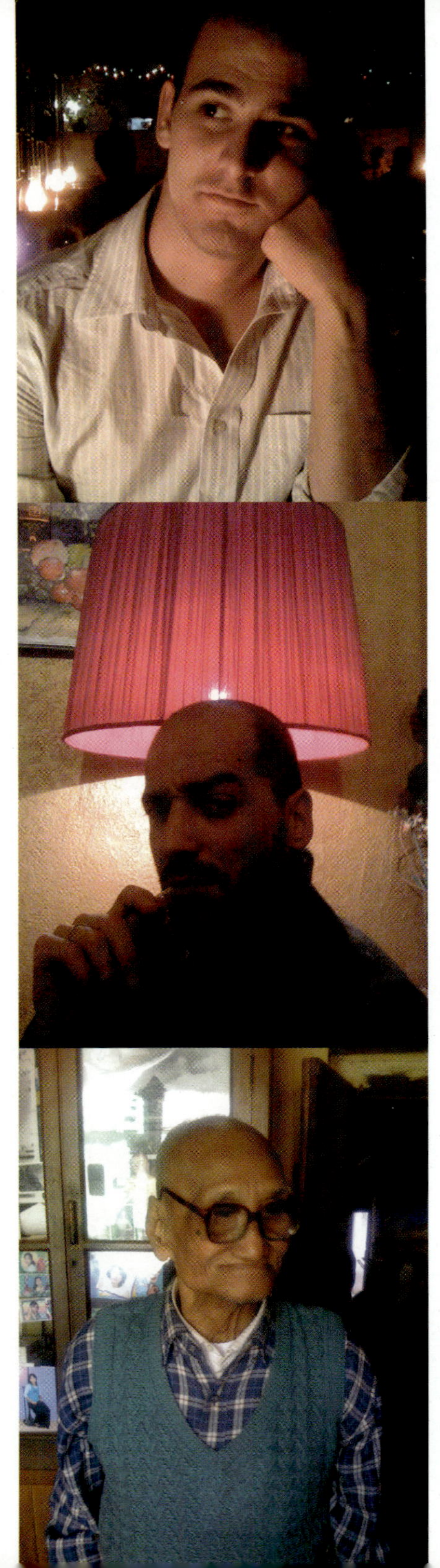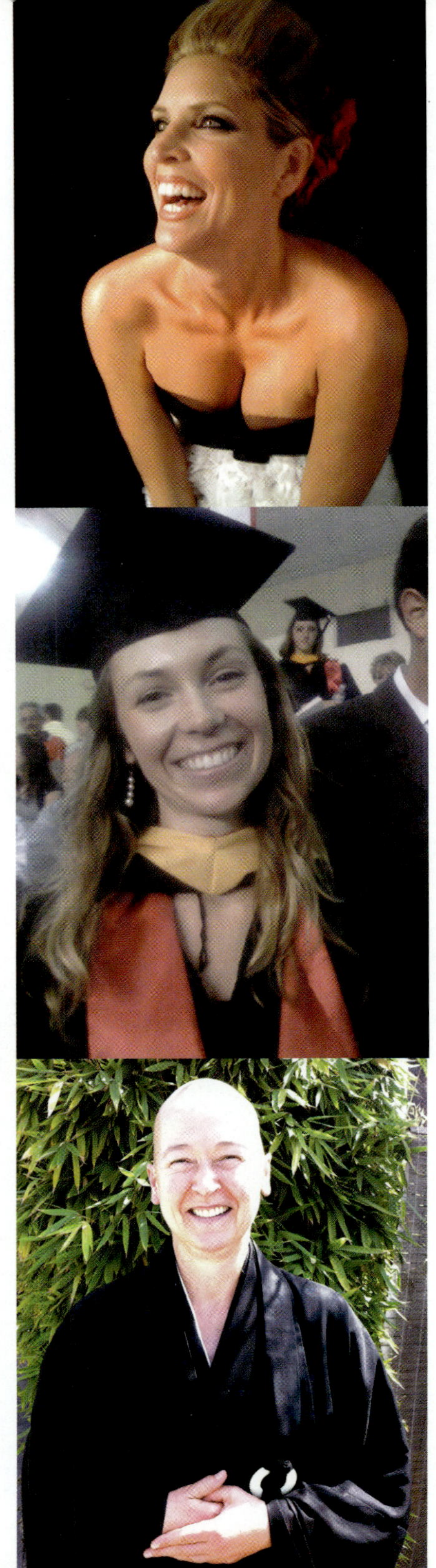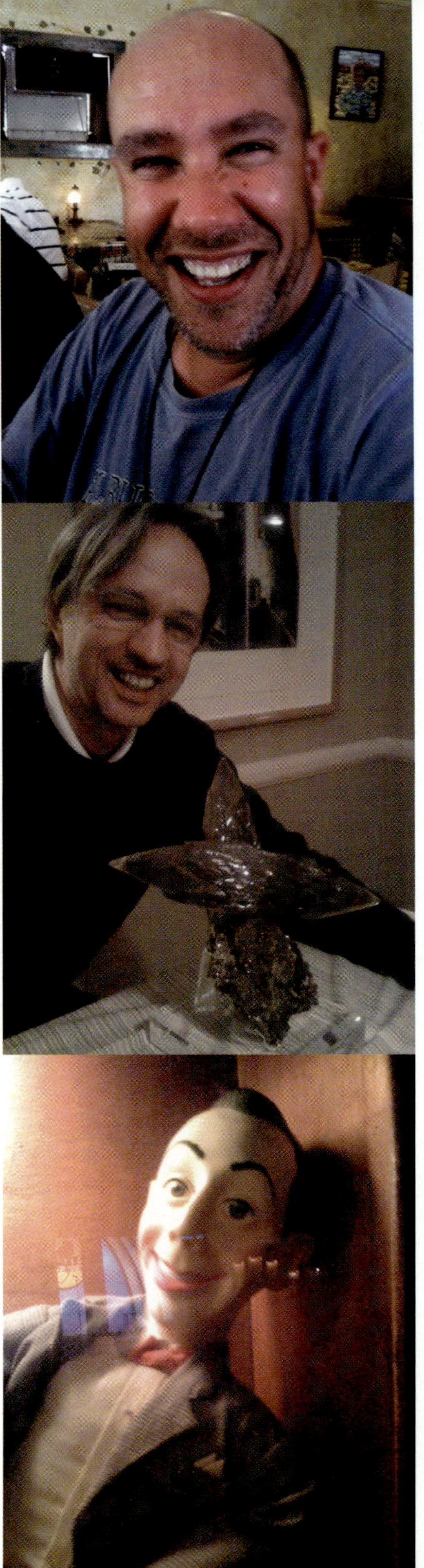

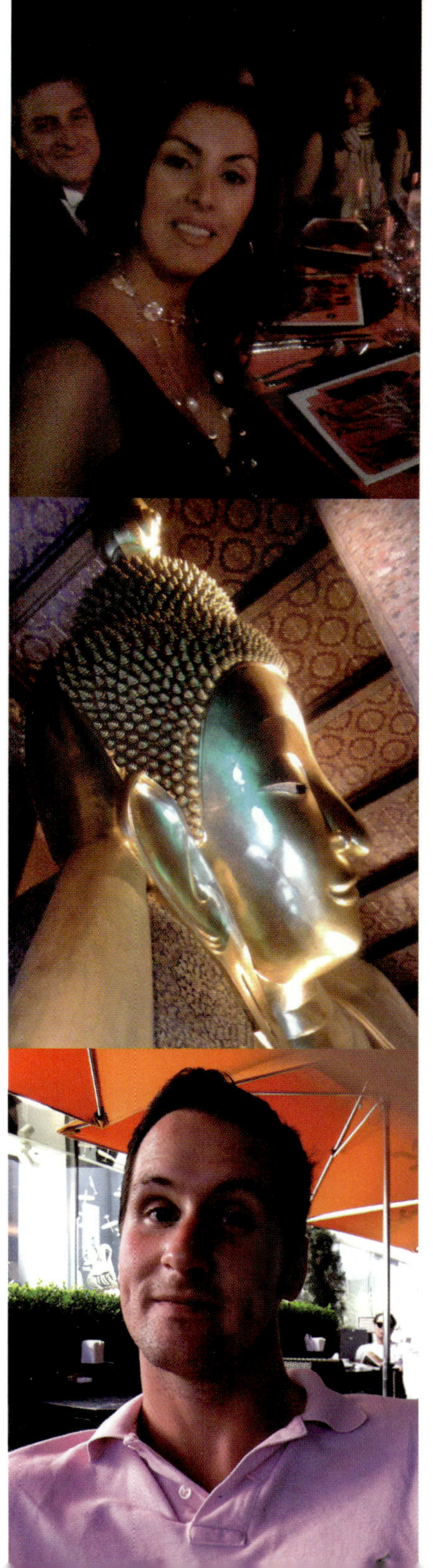

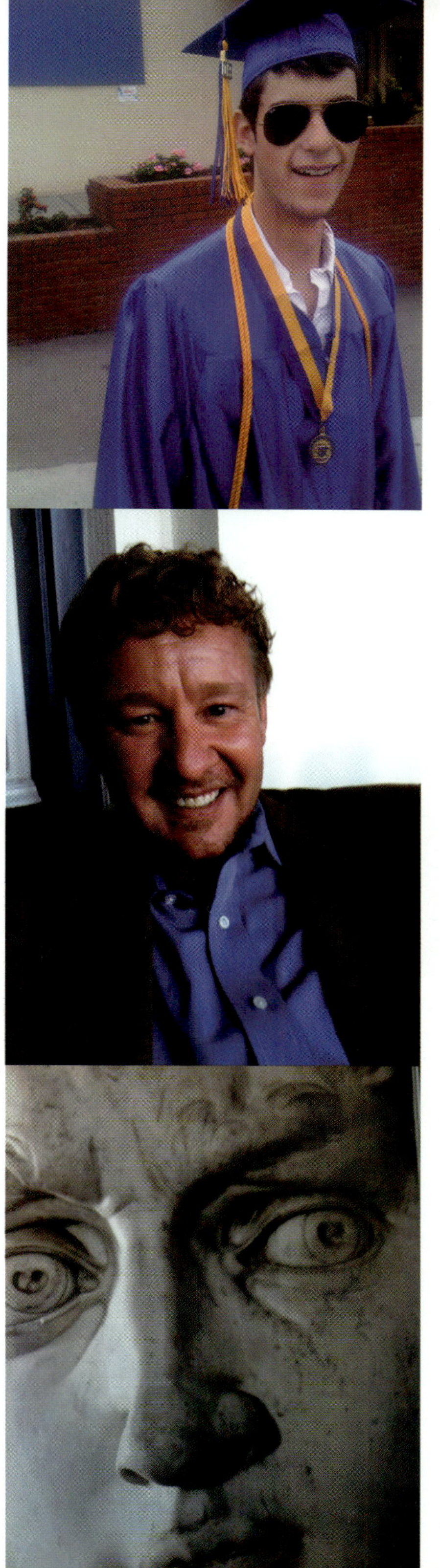

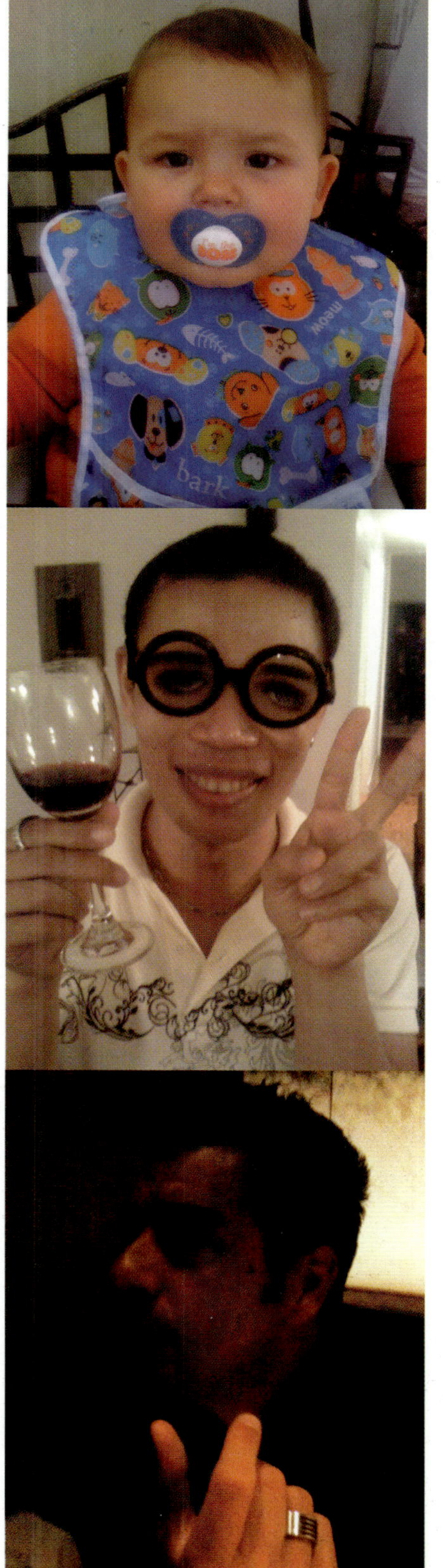

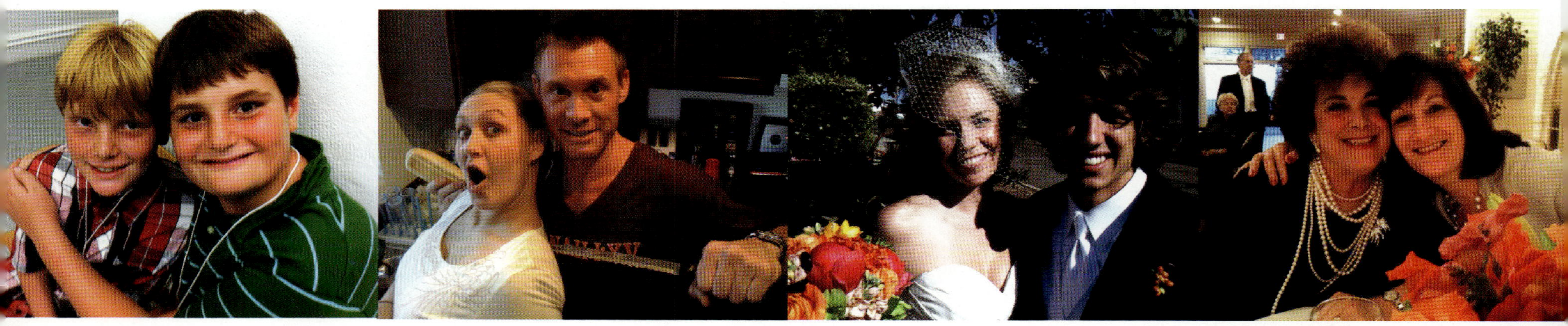

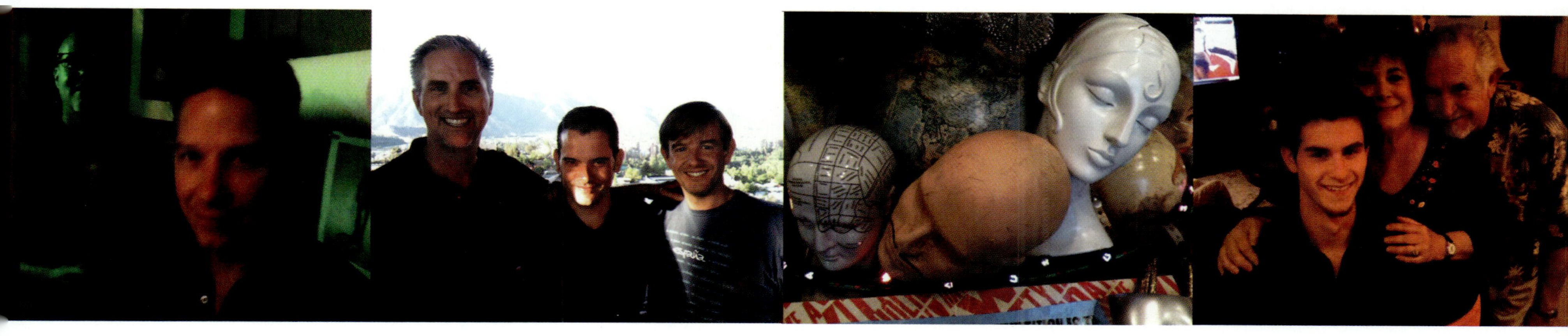

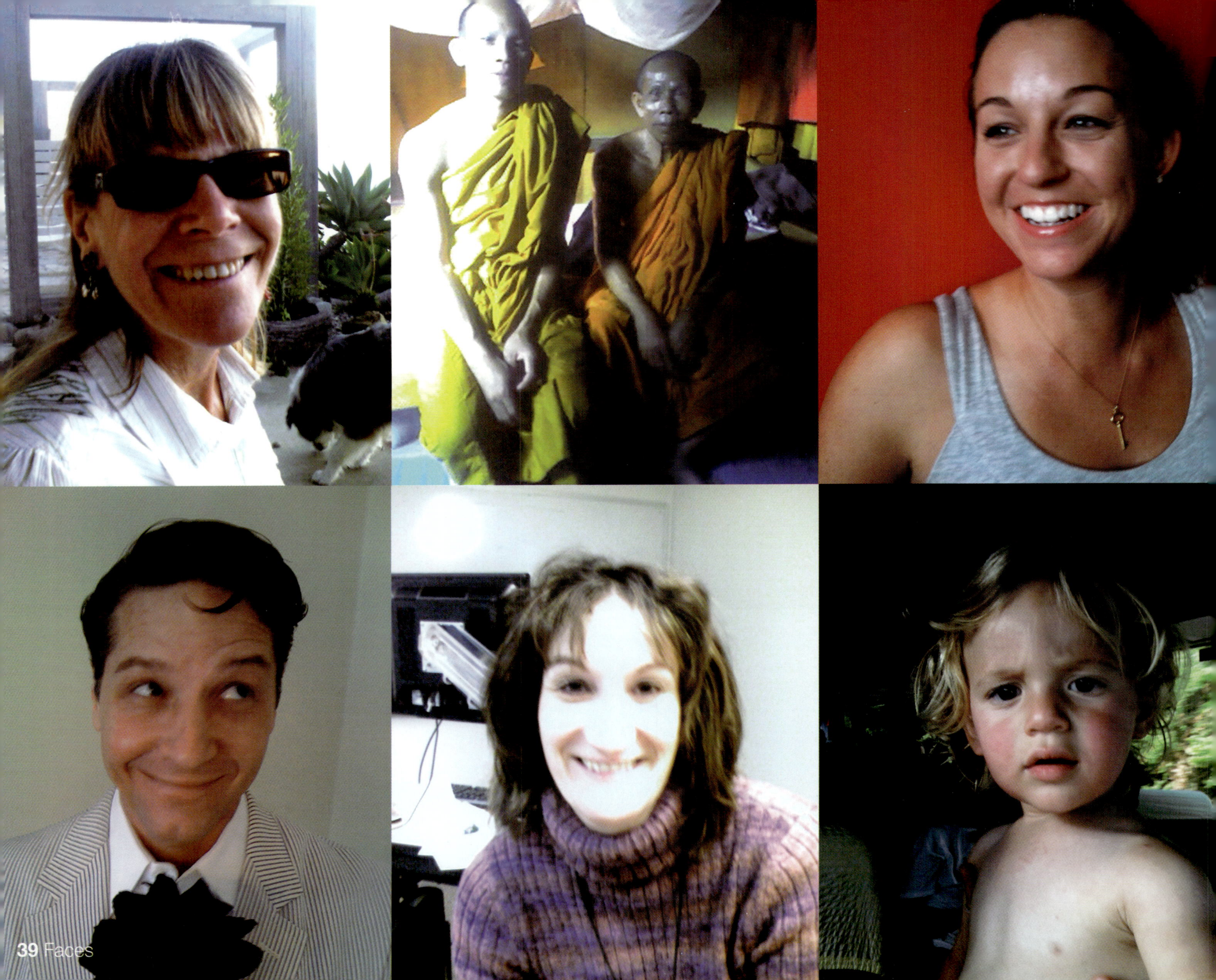

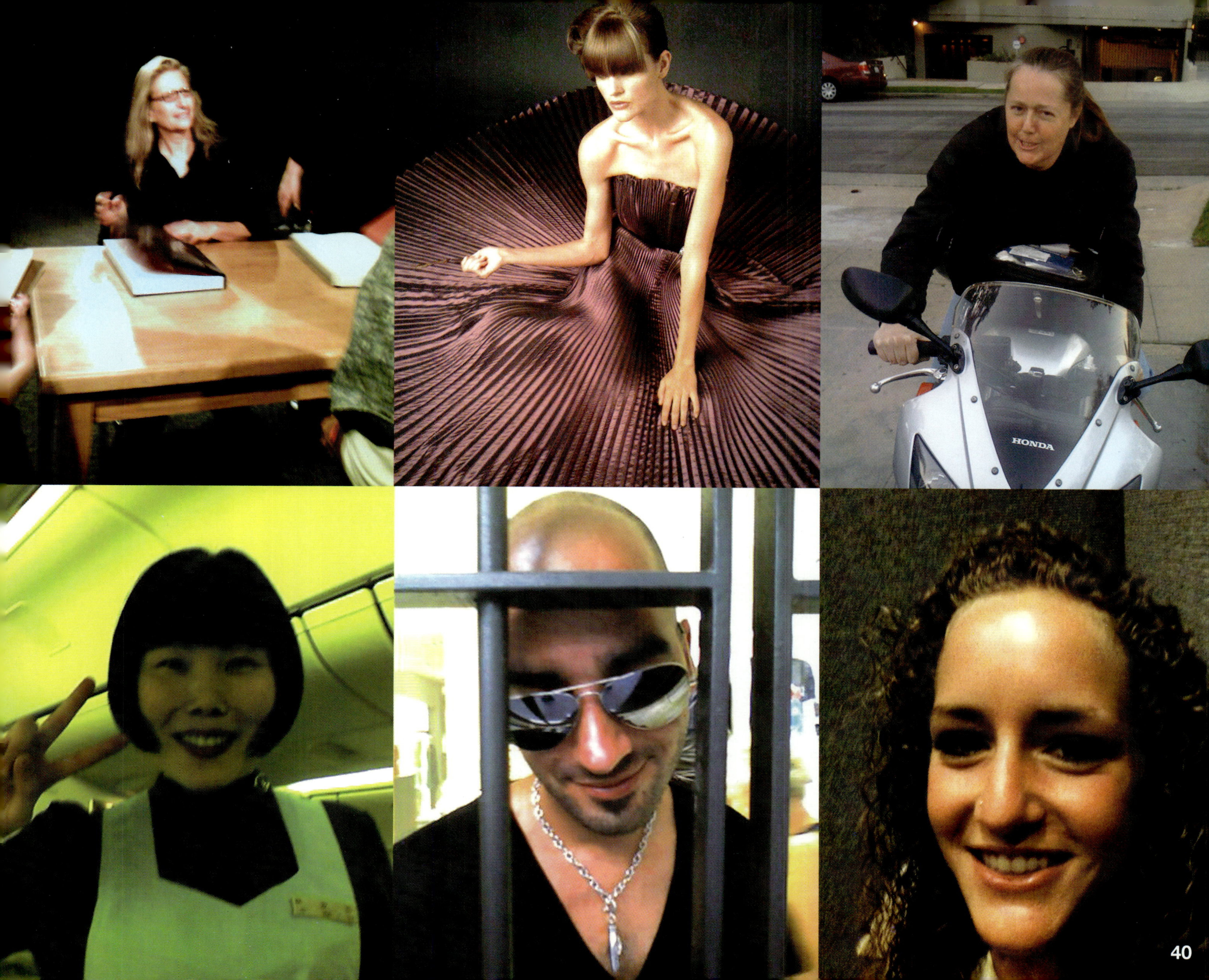

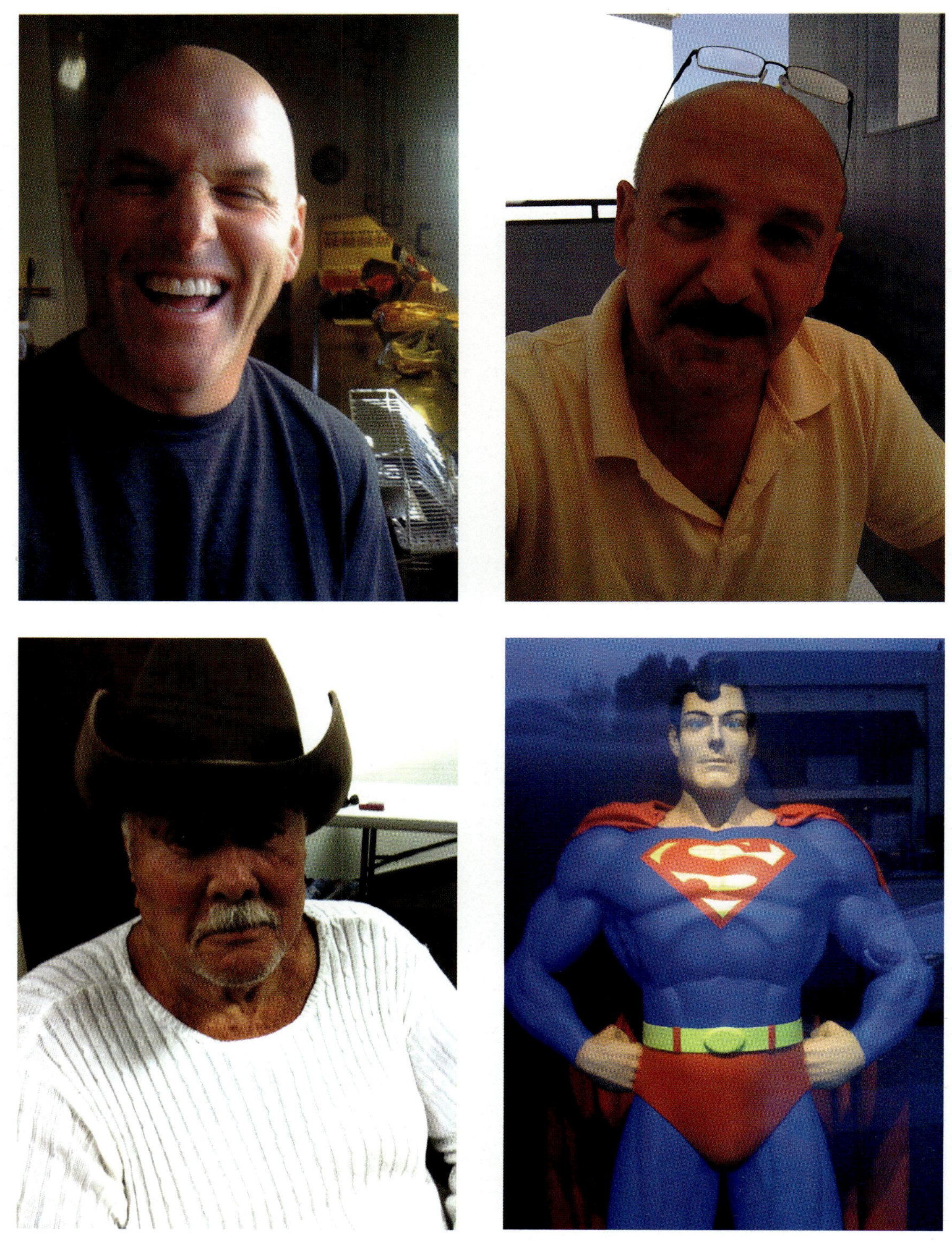

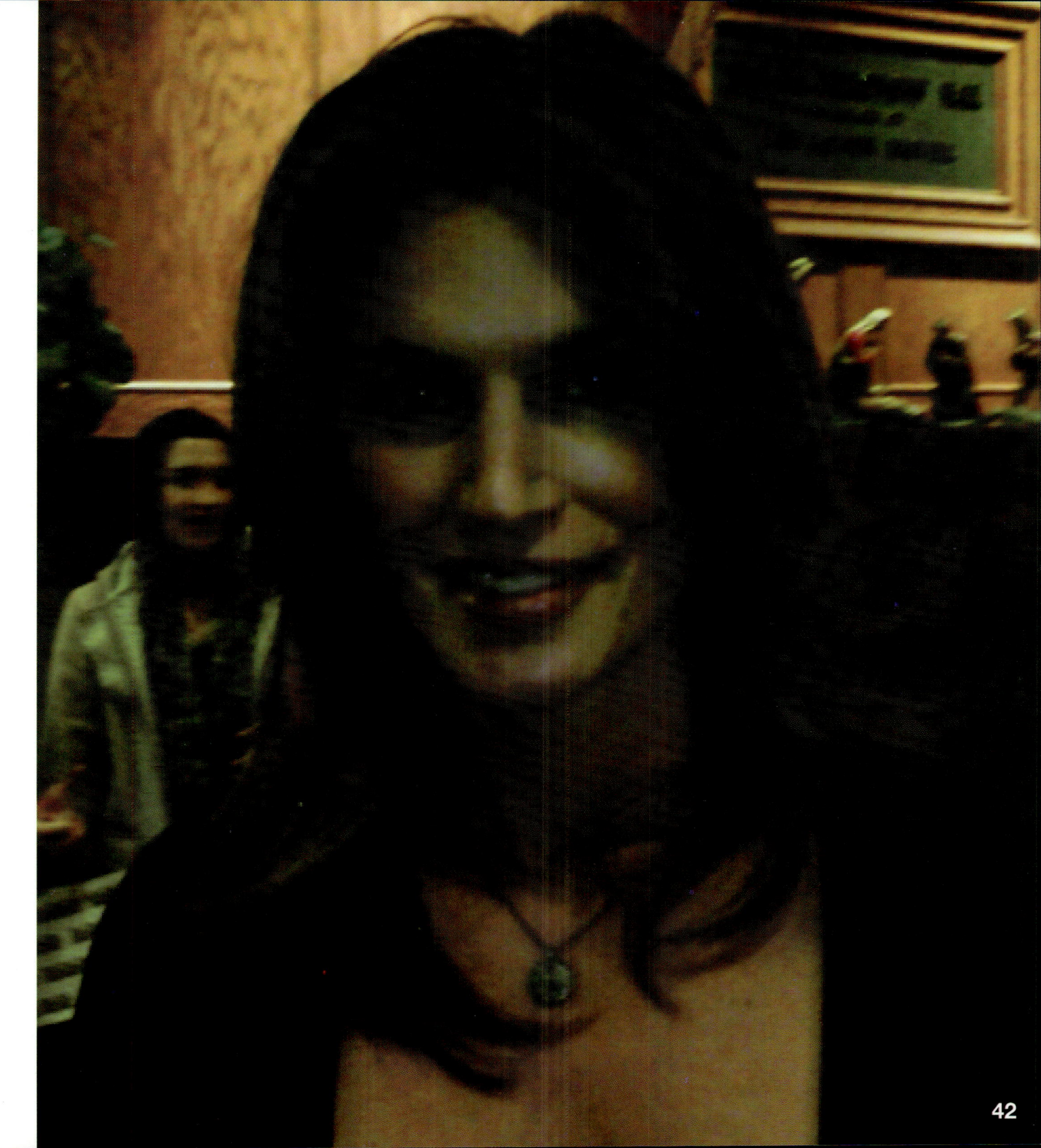

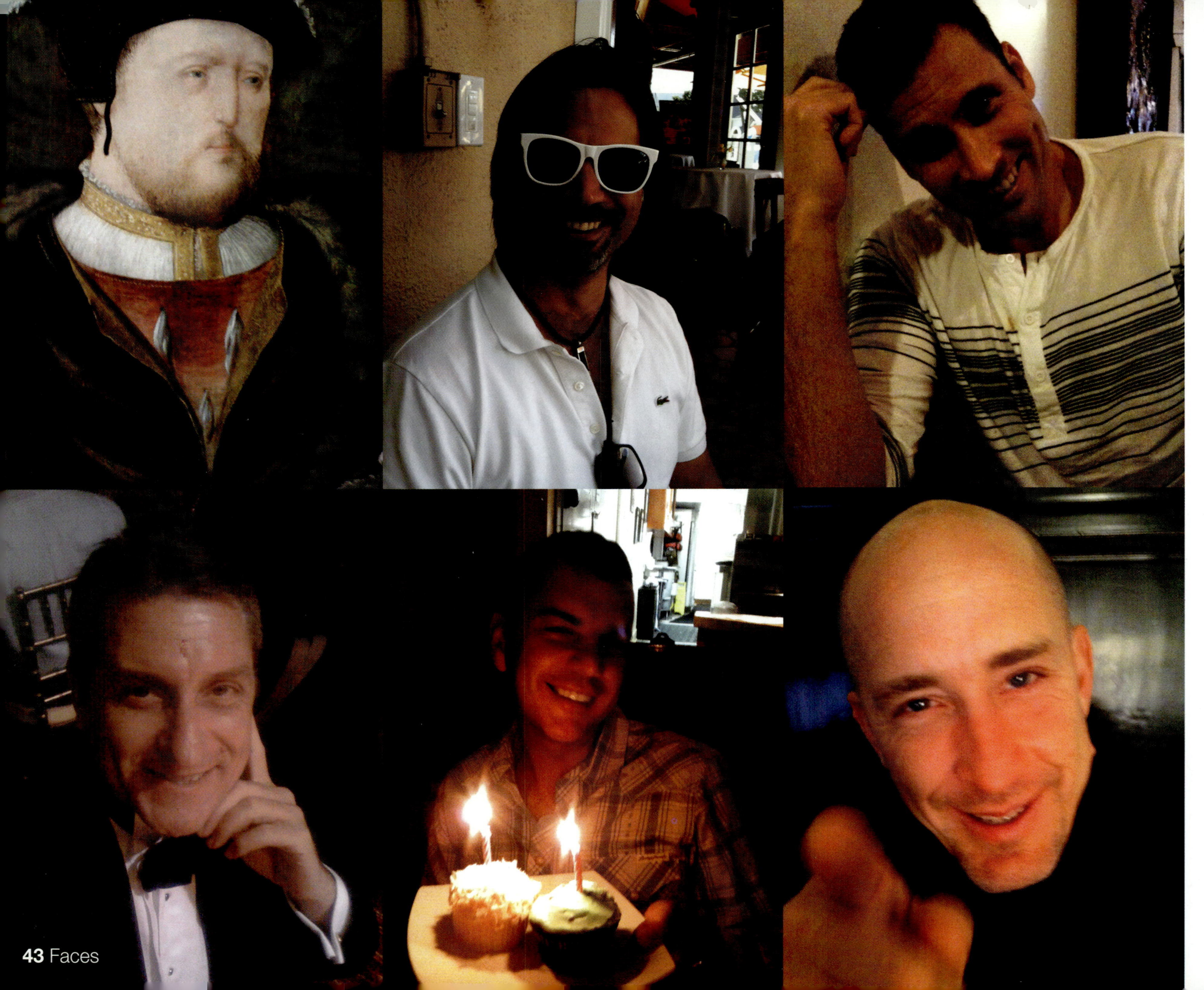

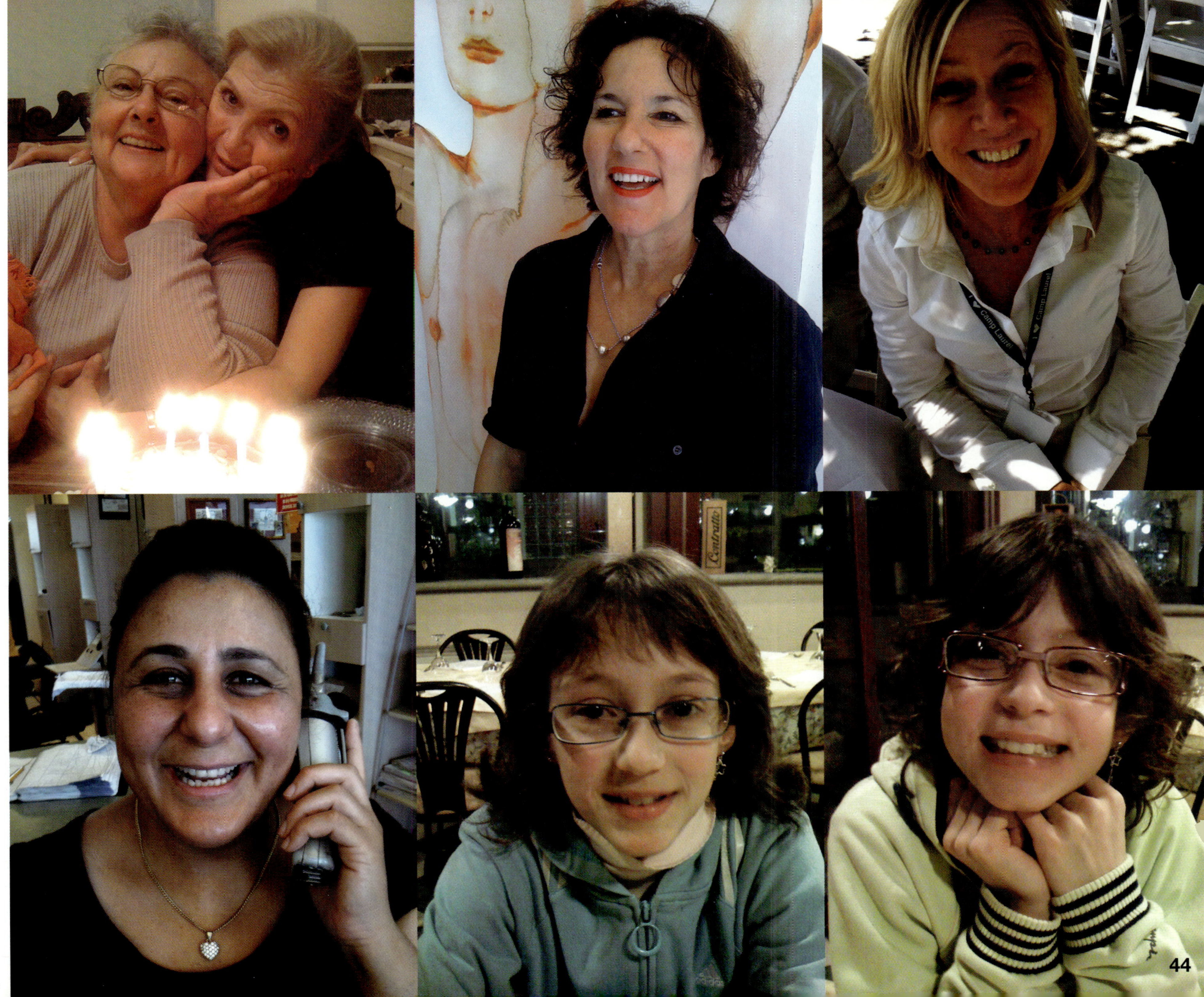

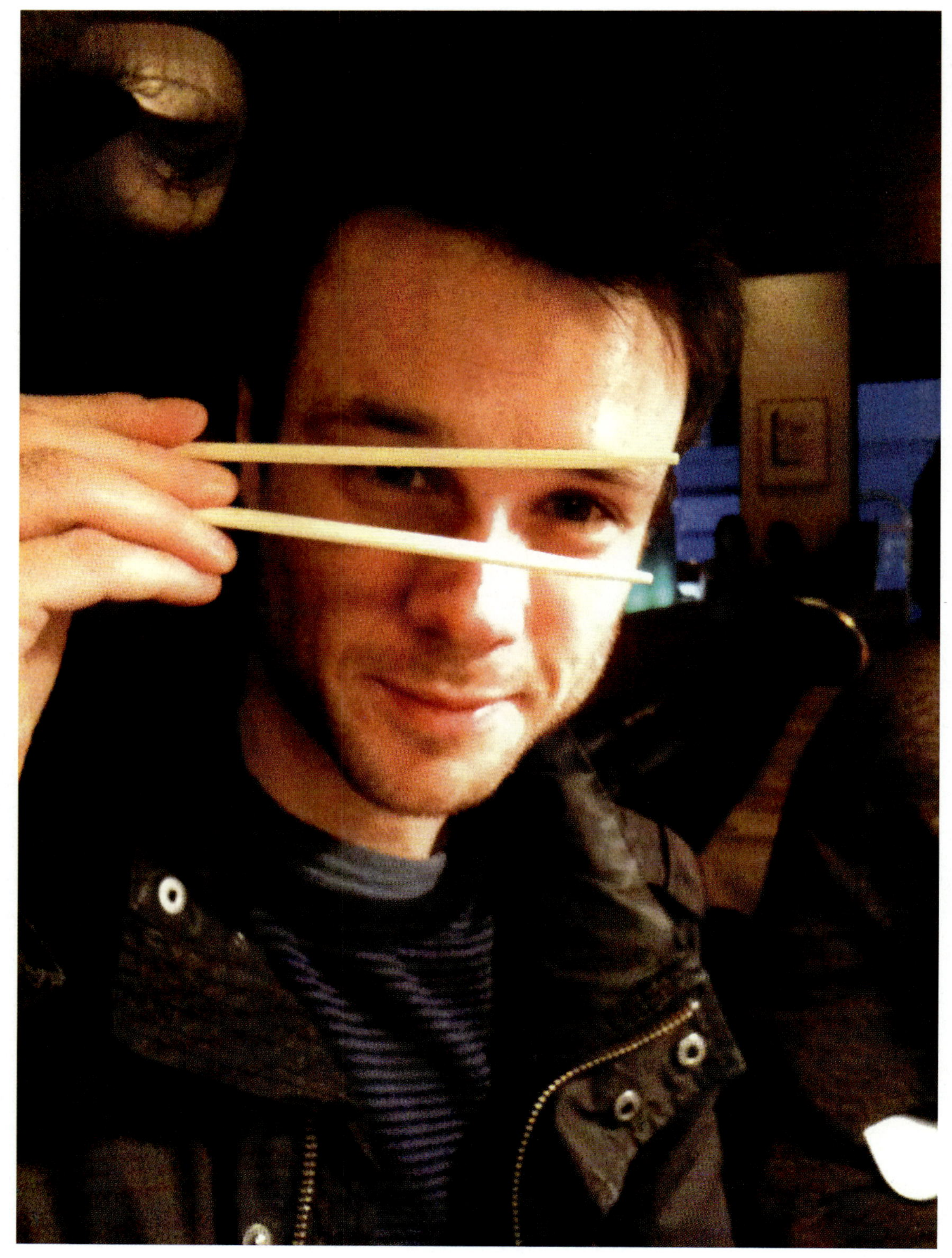

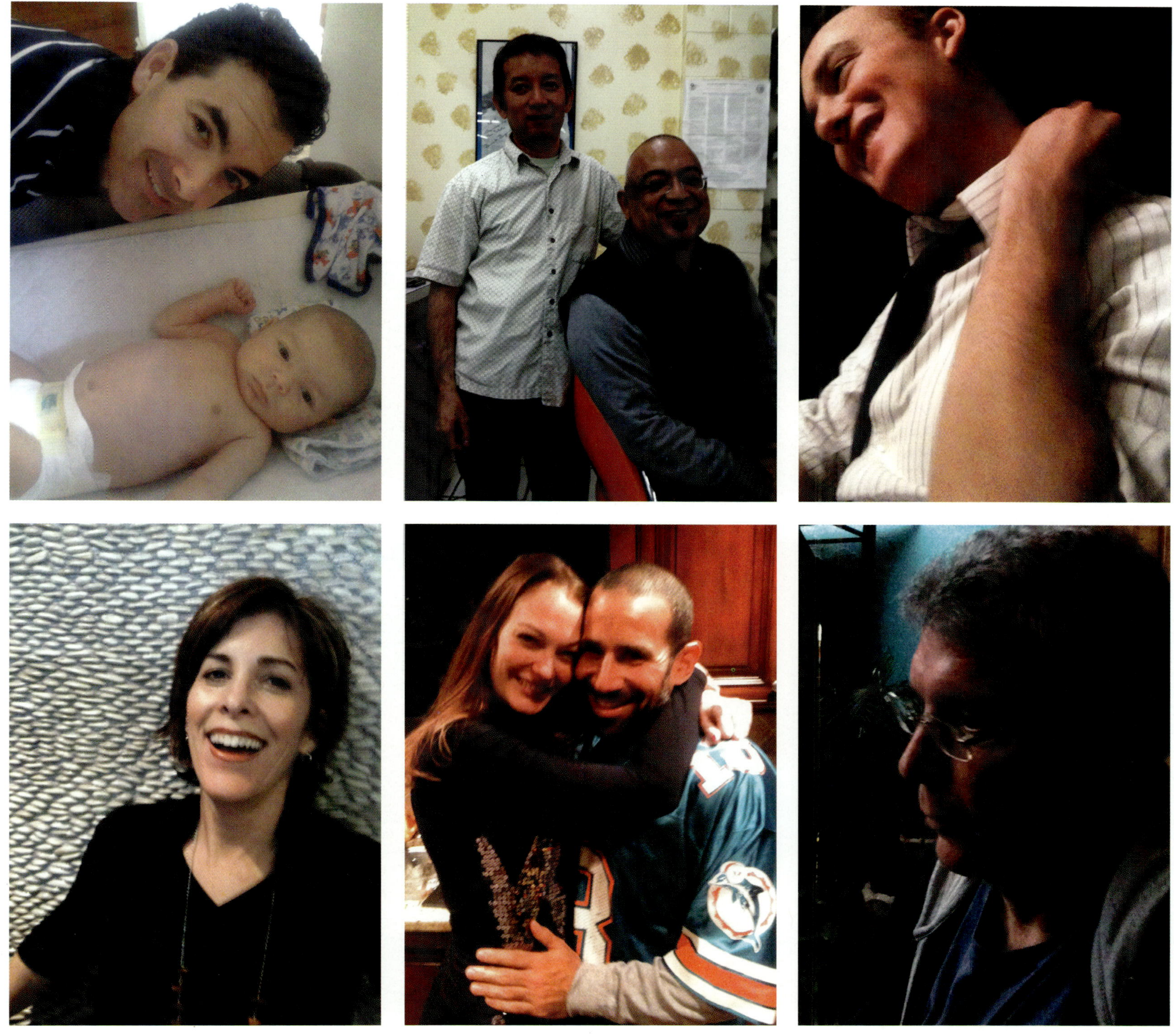

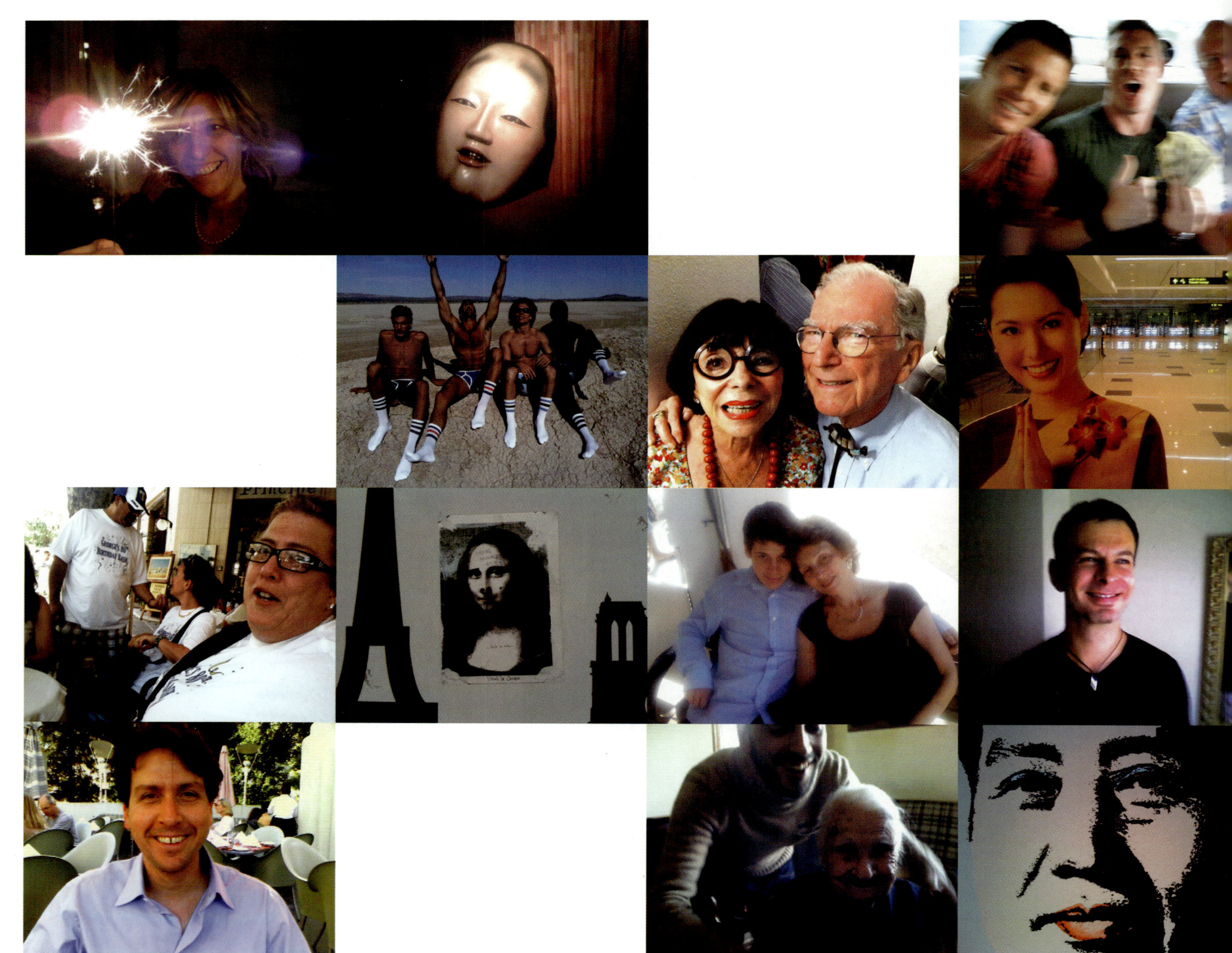

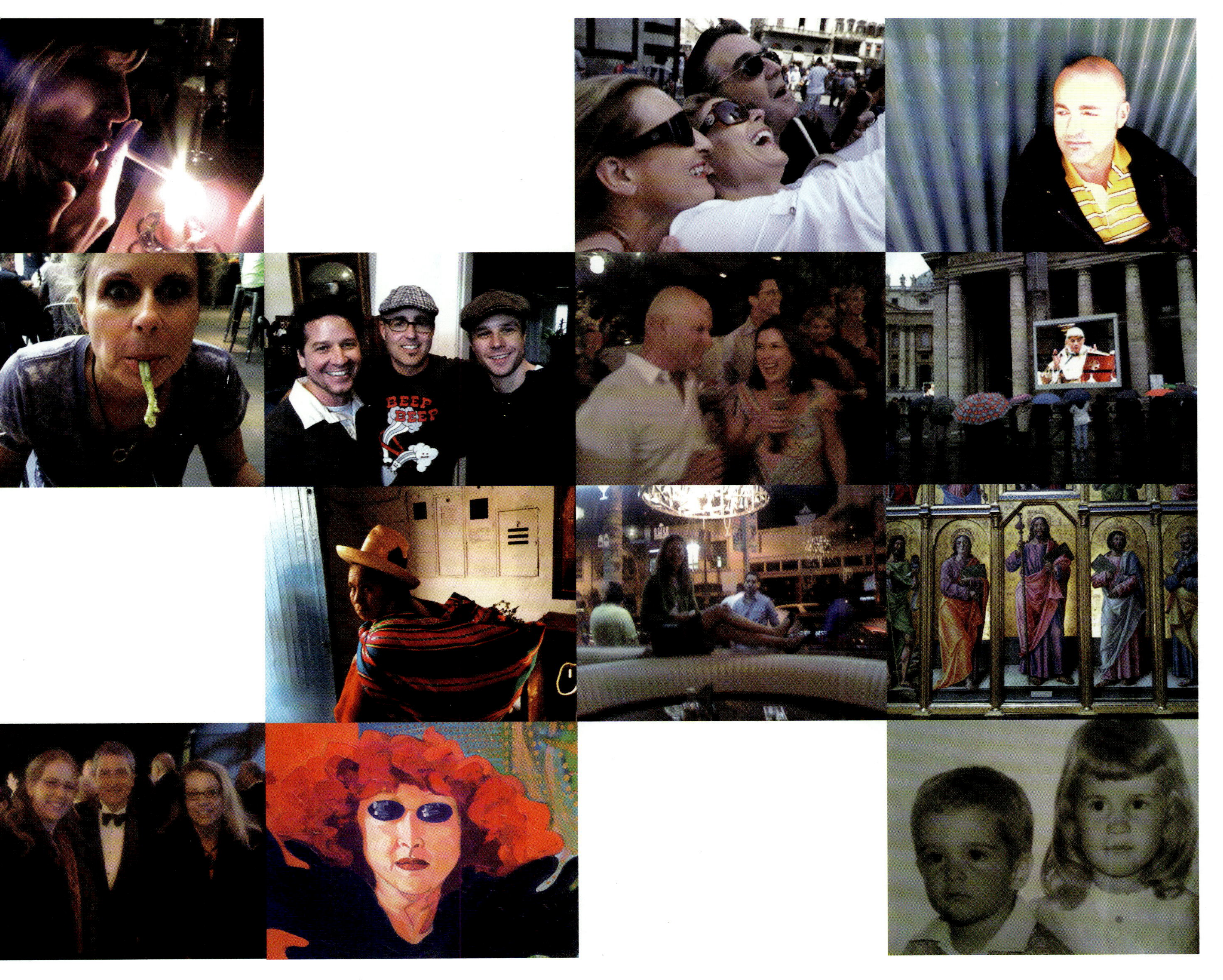

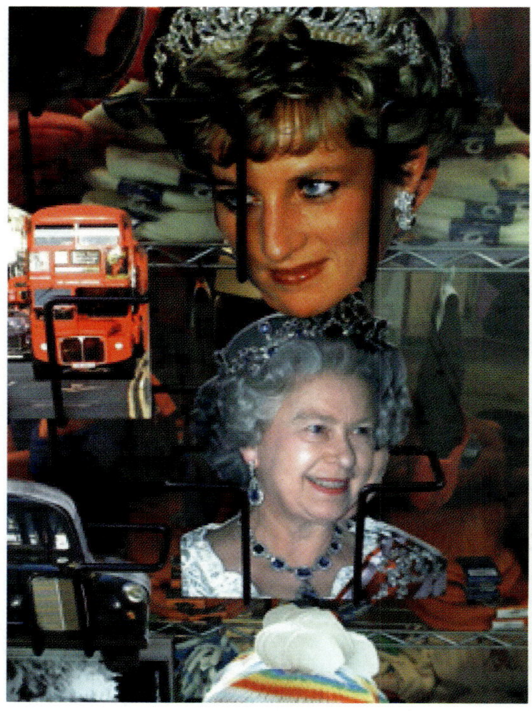

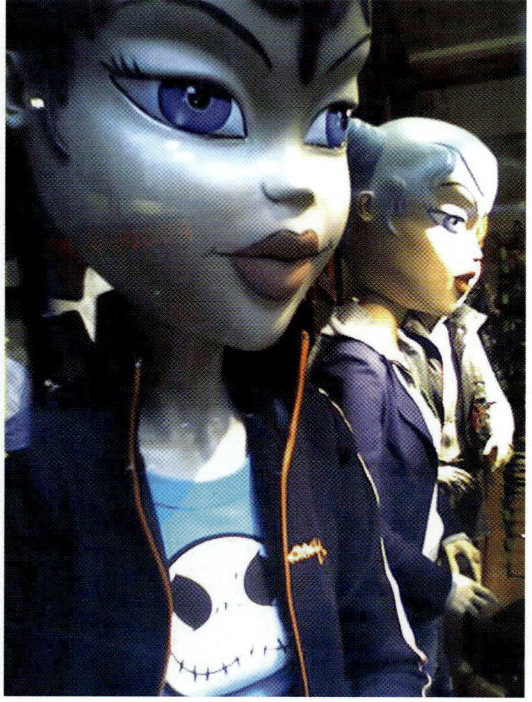

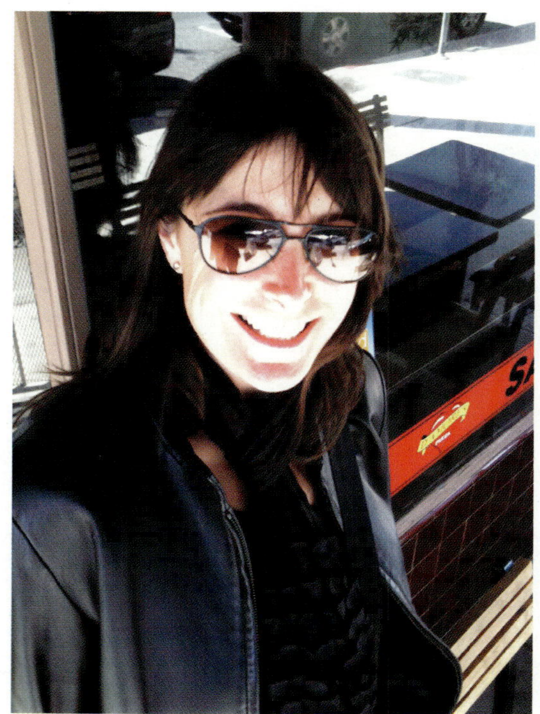

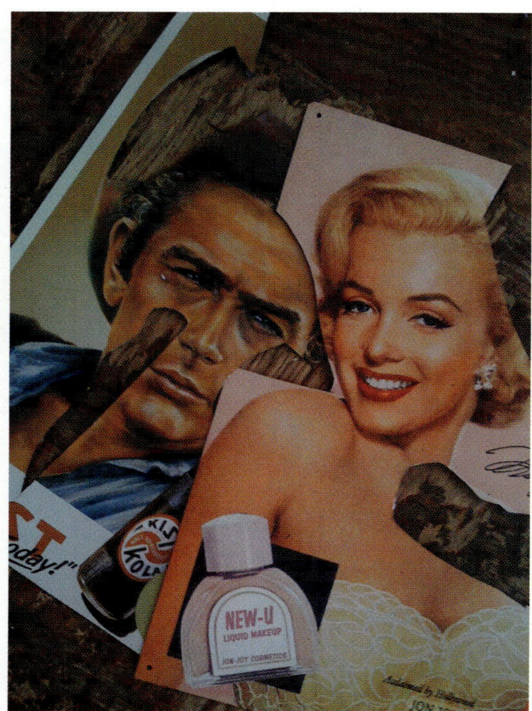
NEW-U
LIQUID MAKEUP
JON-JOY COSMETICS

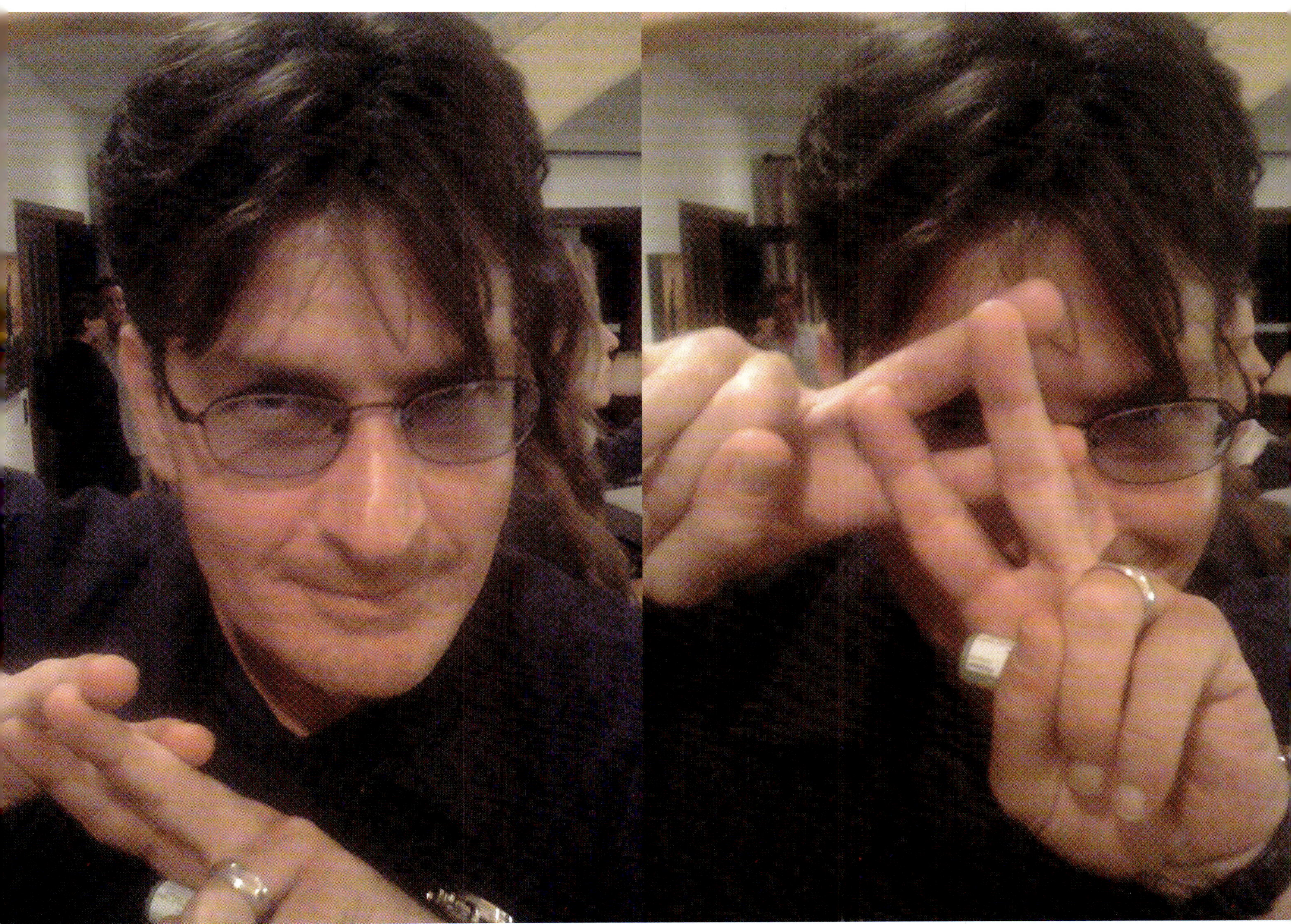

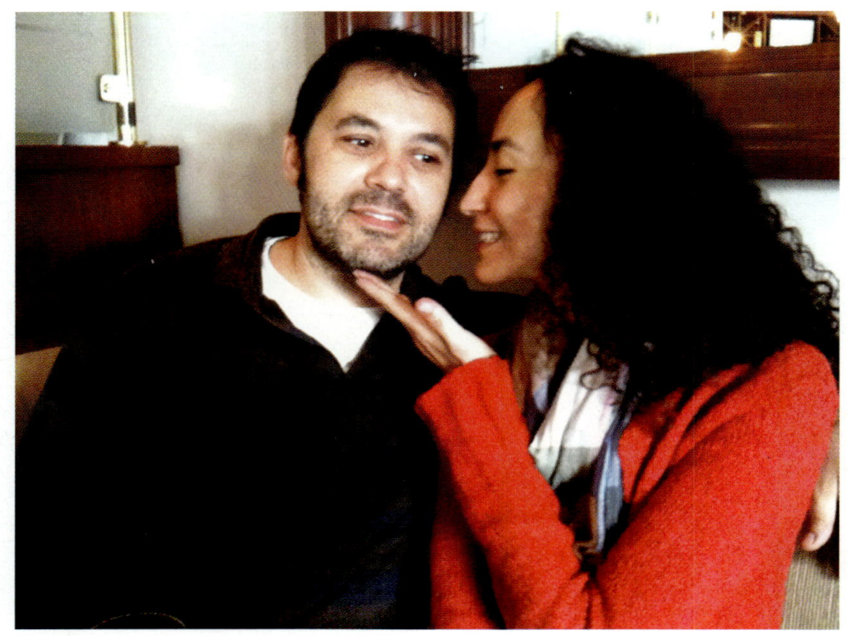
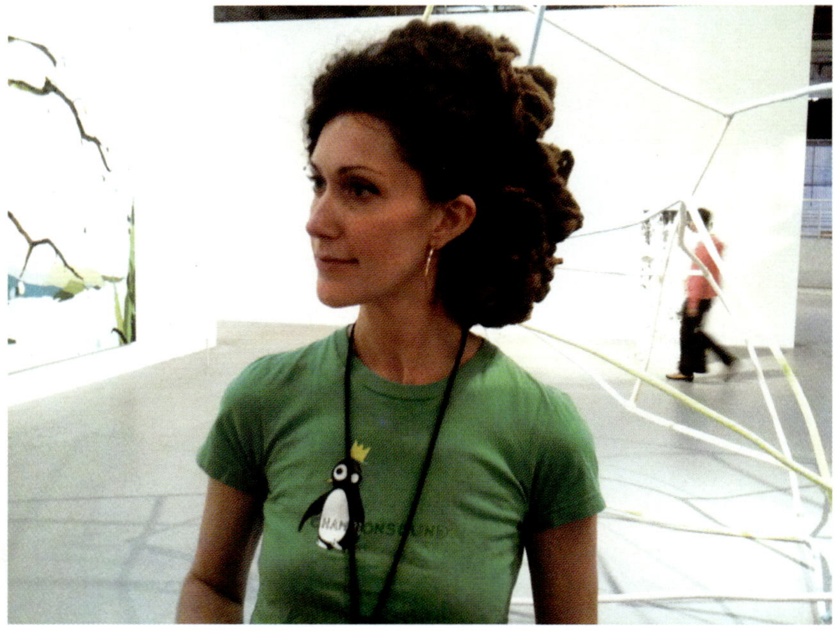
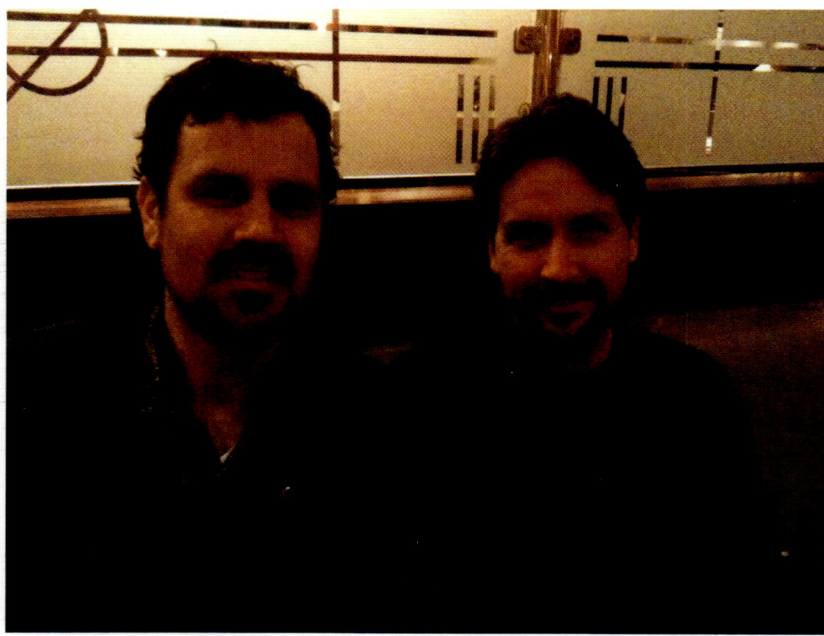
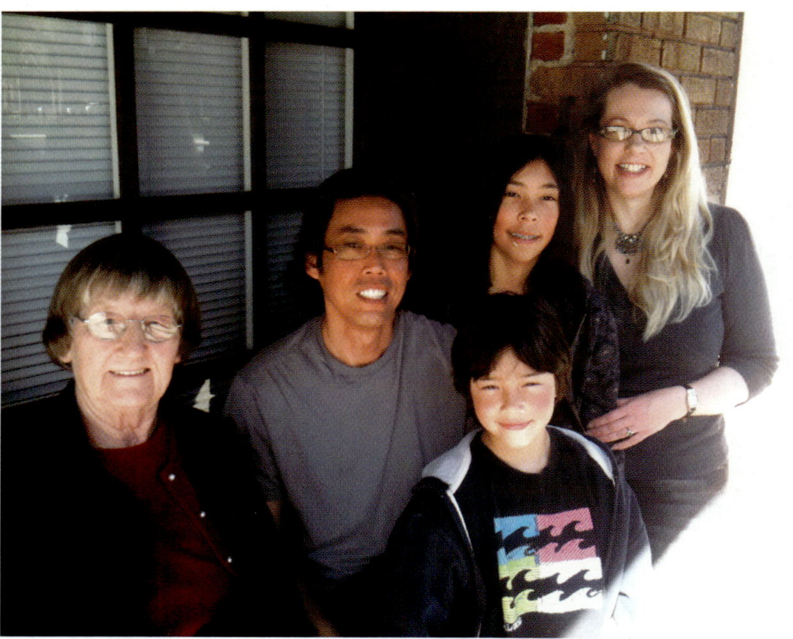

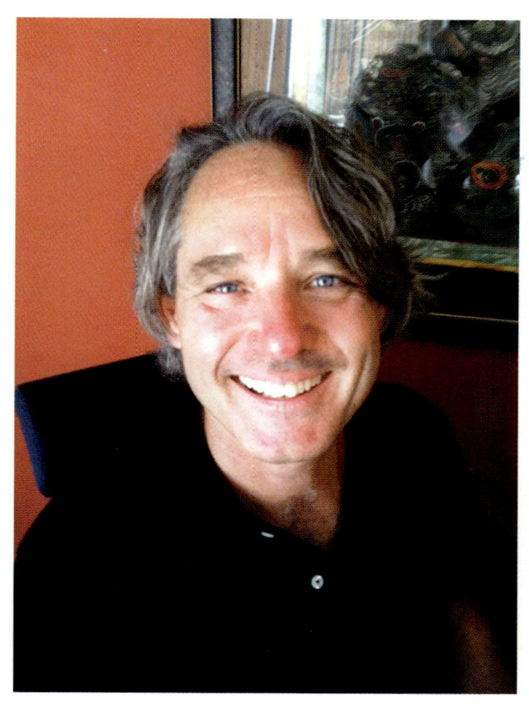
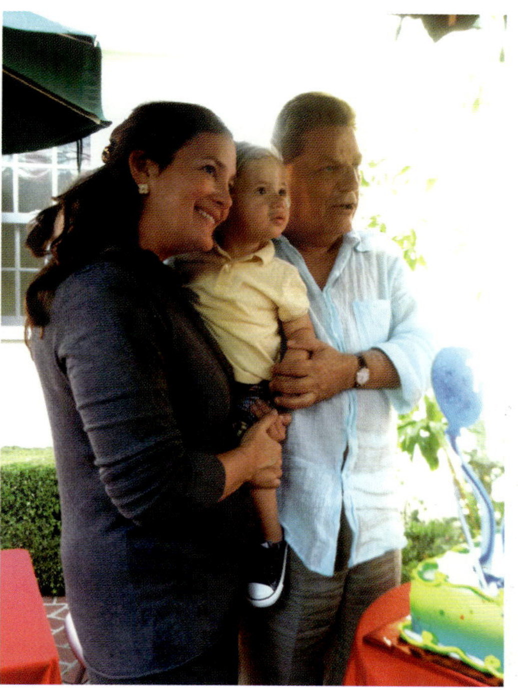
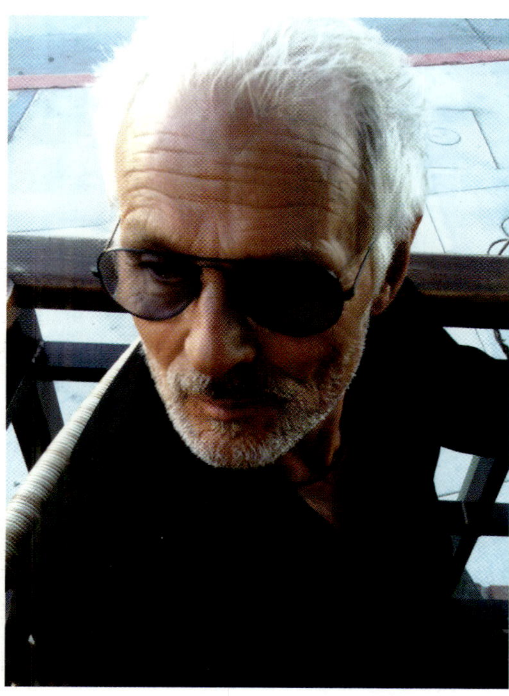
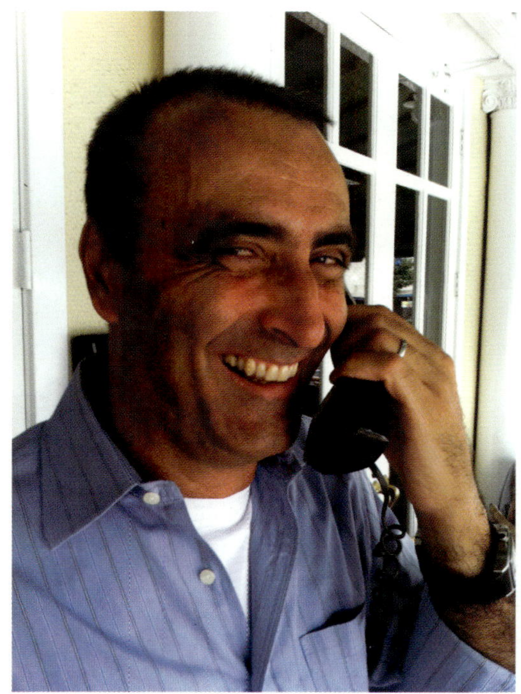
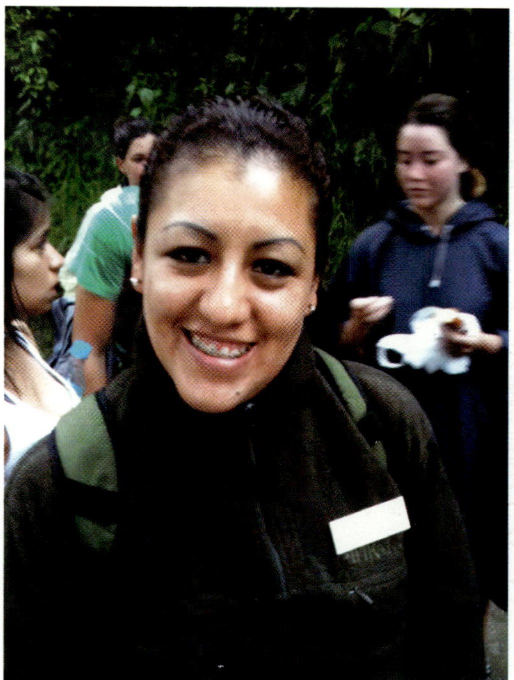
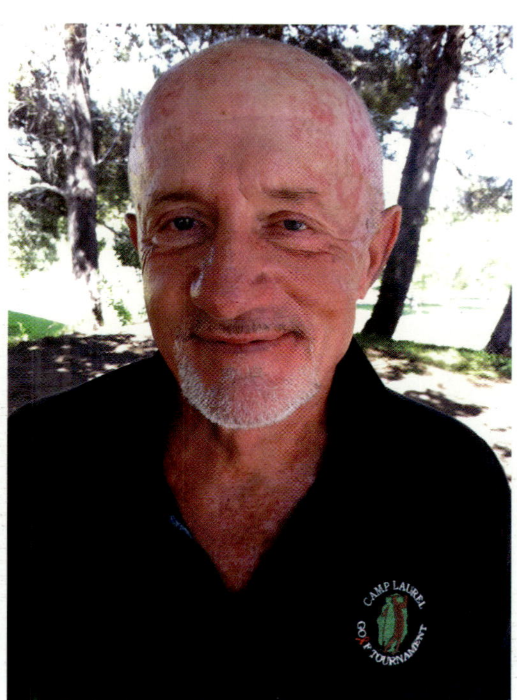

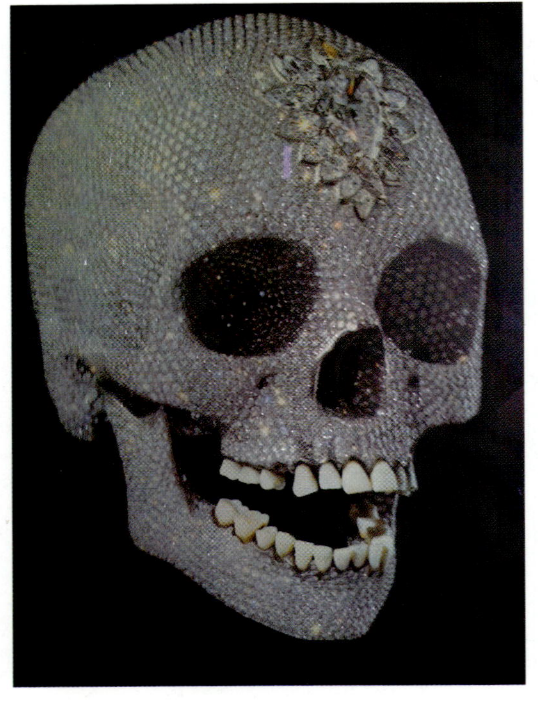 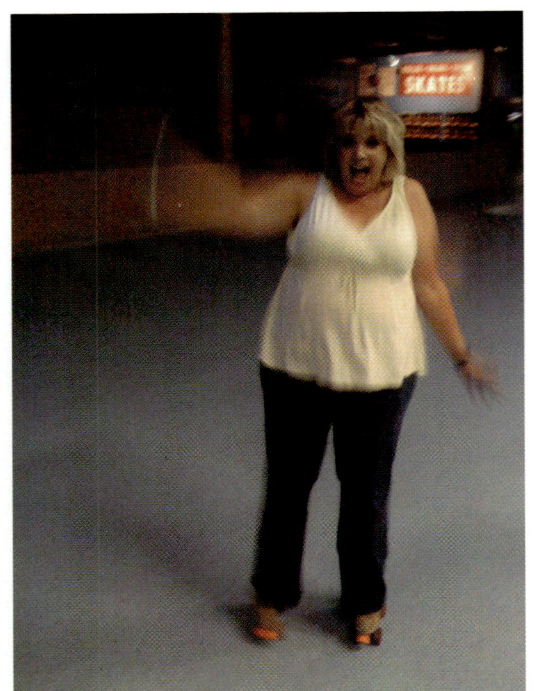 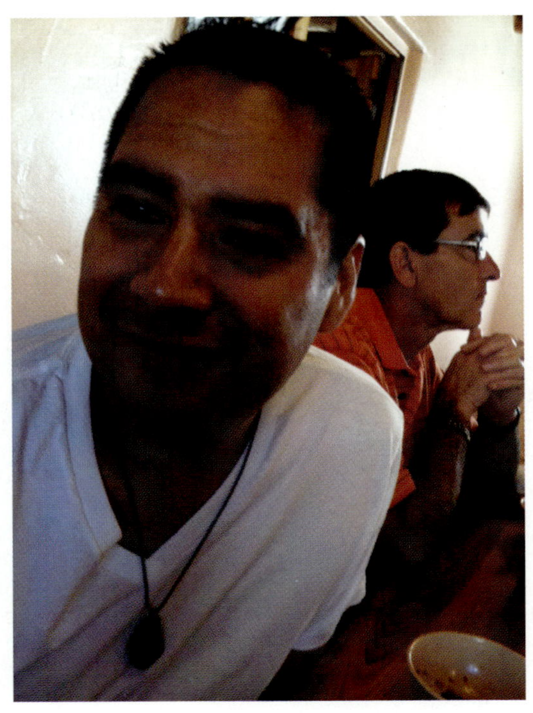

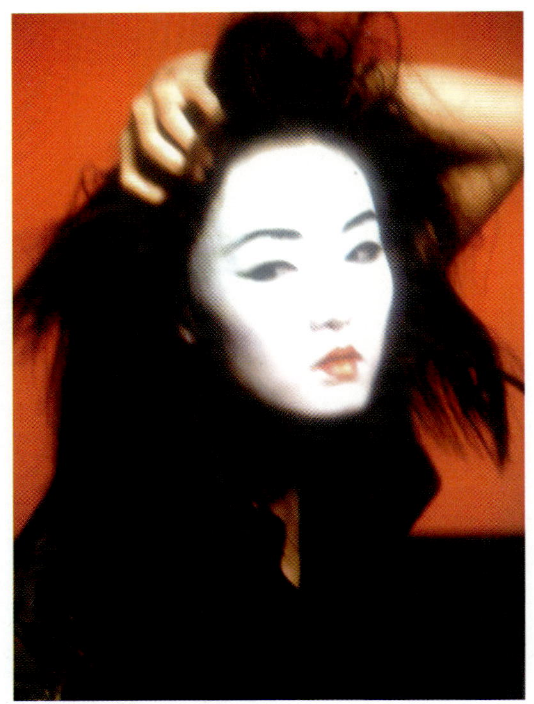 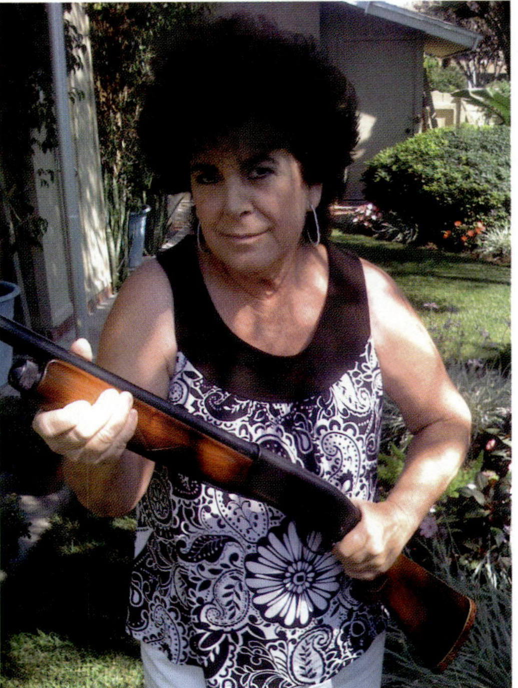 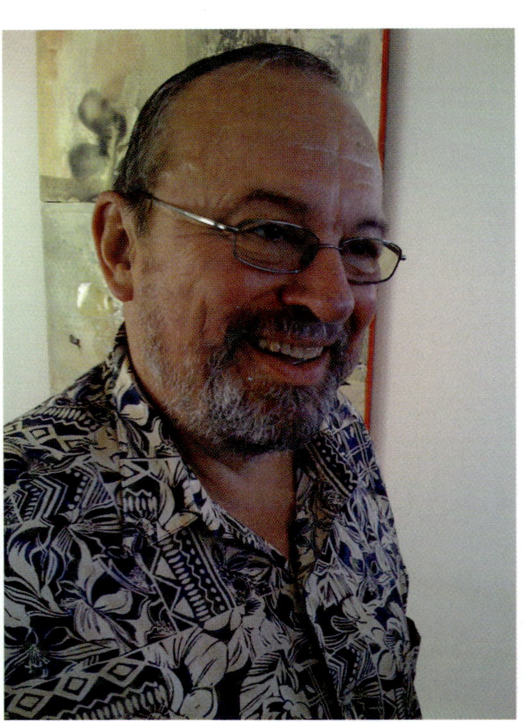

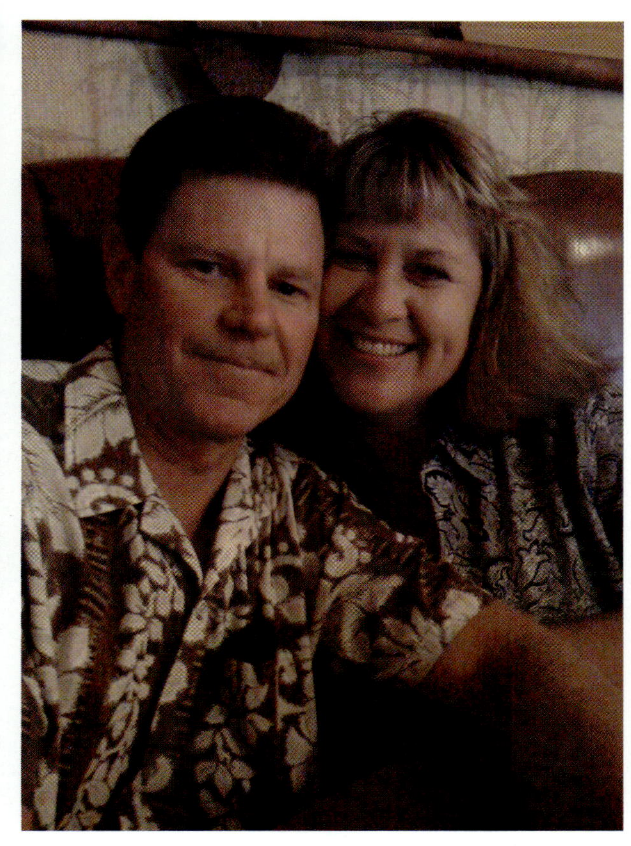
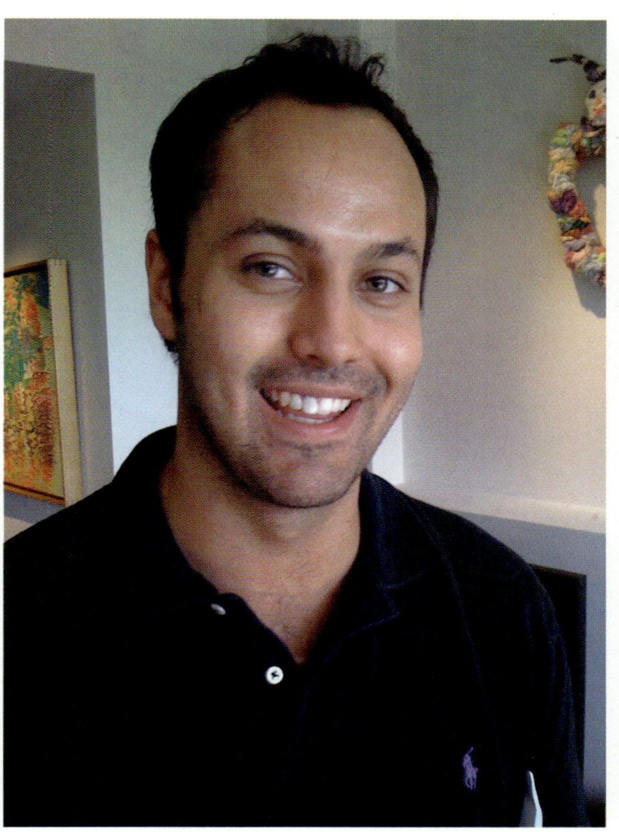
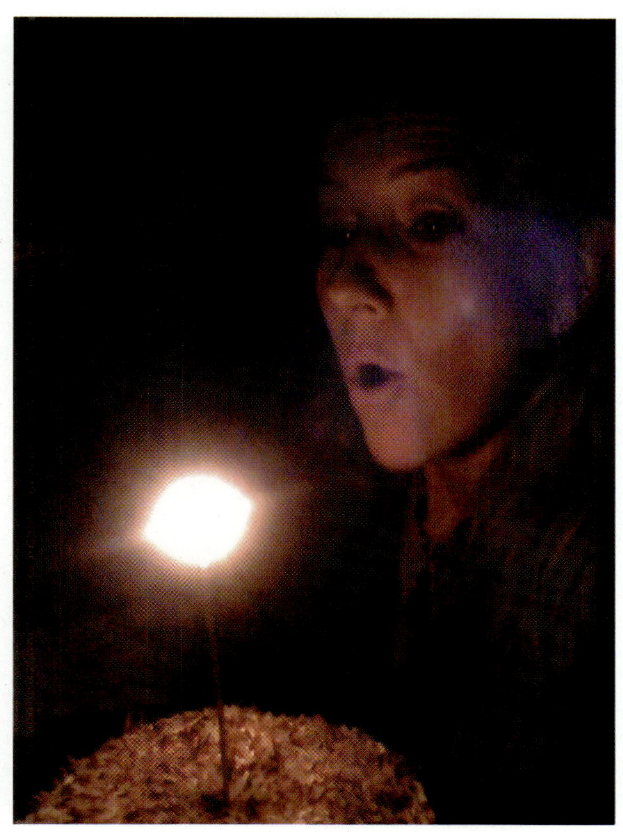

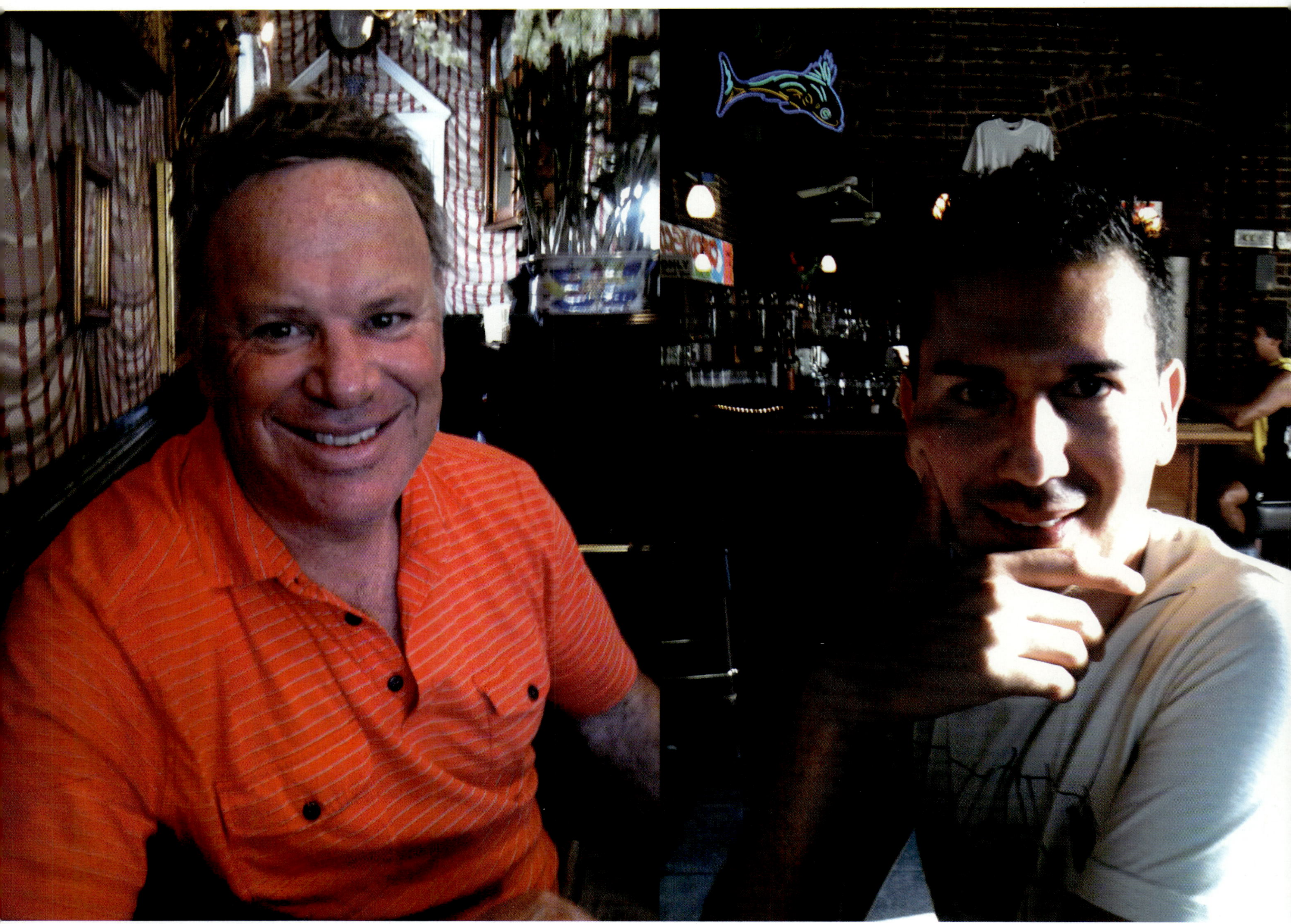

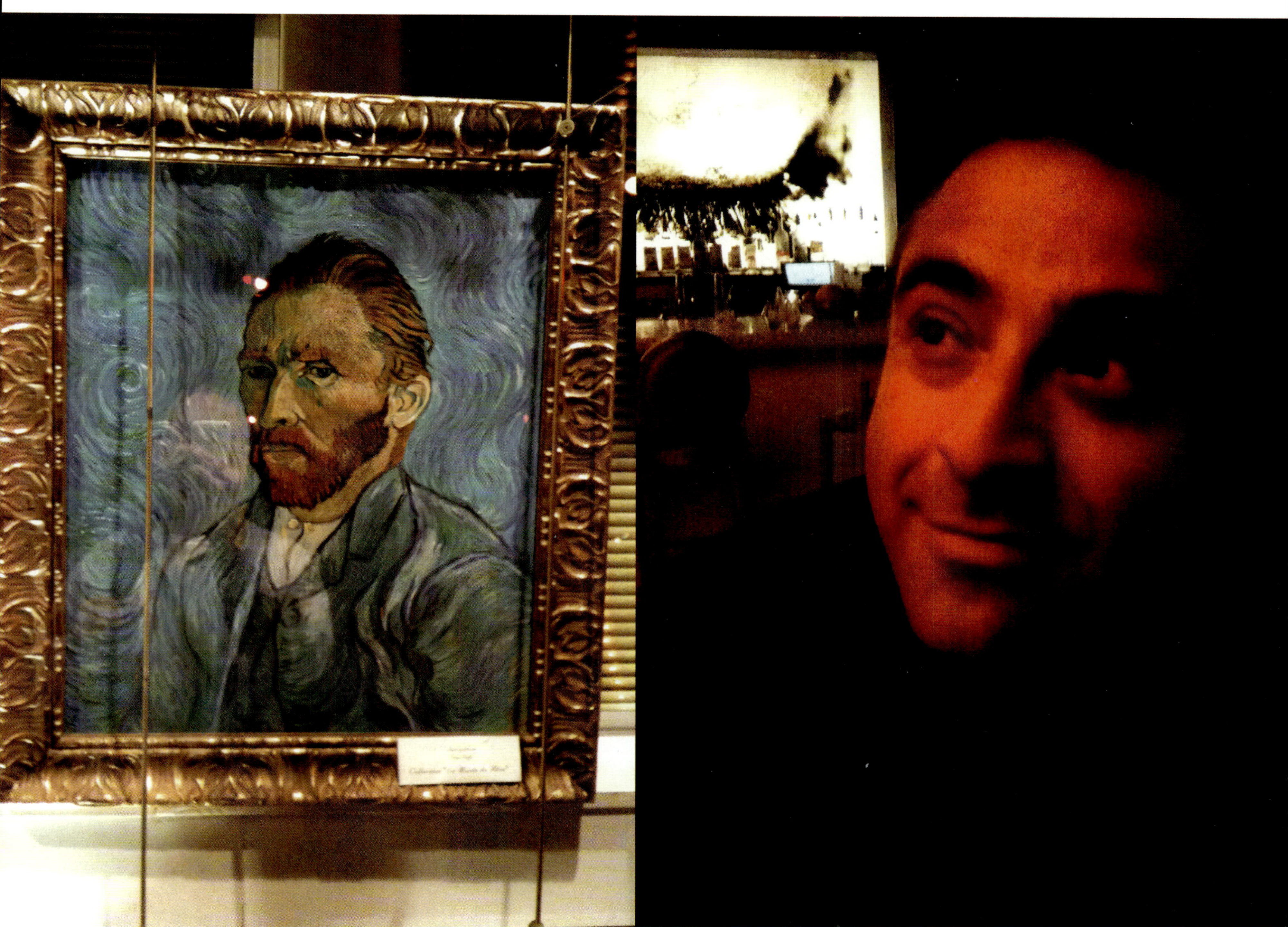

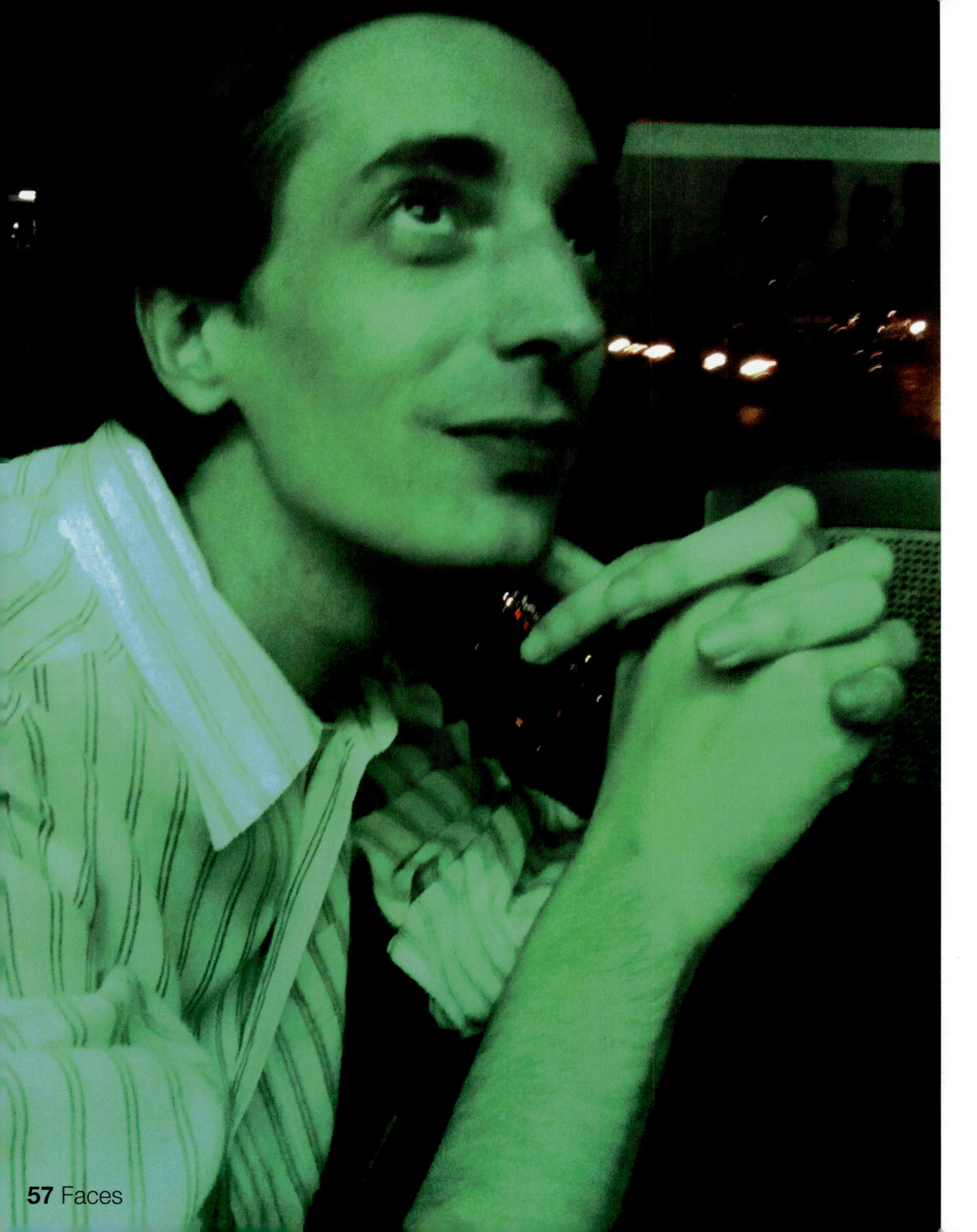

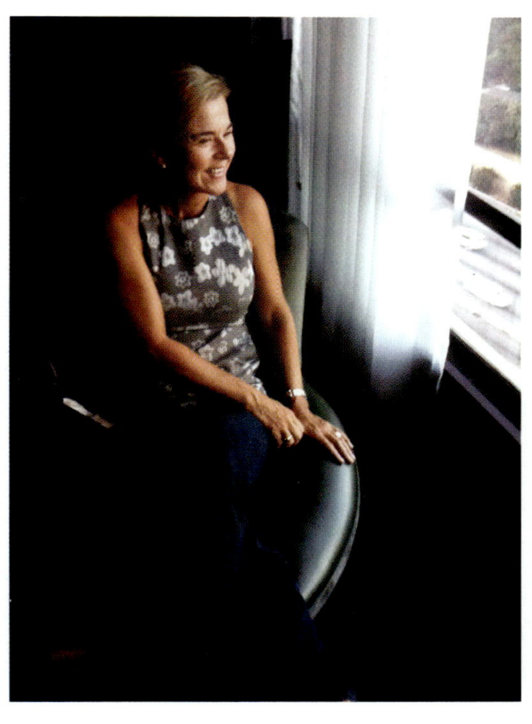
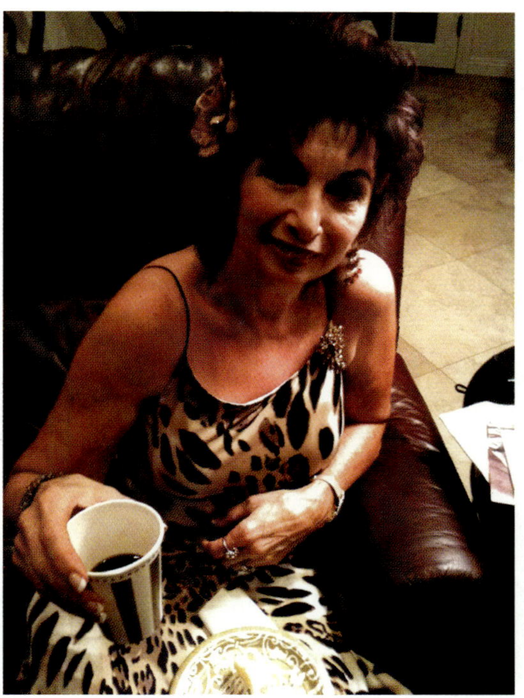
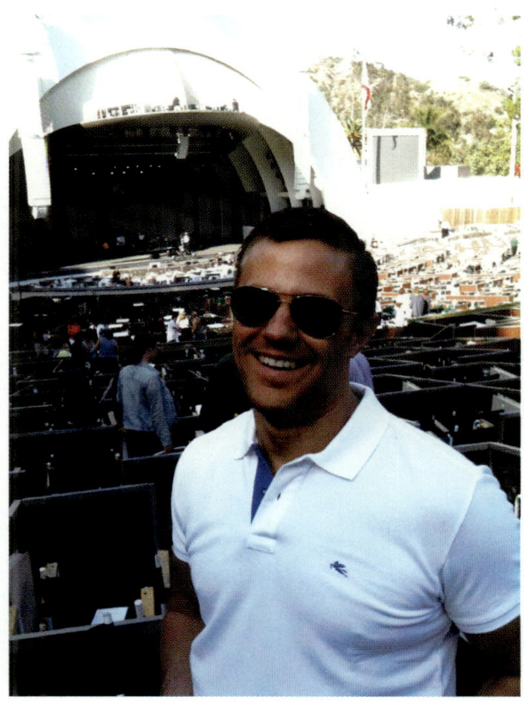
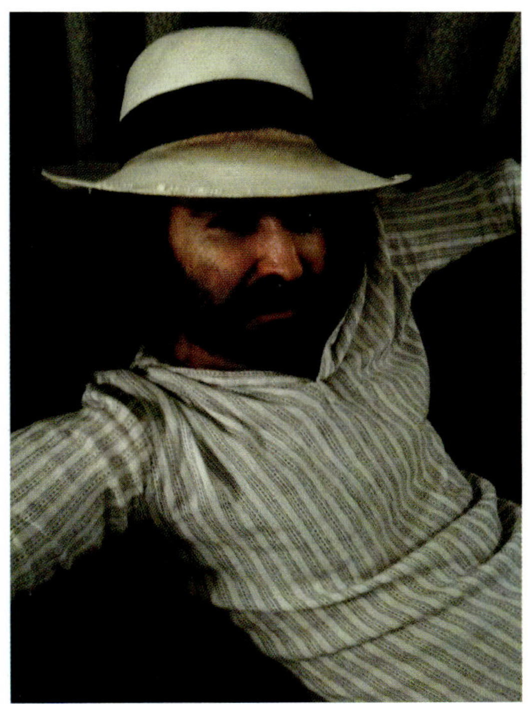
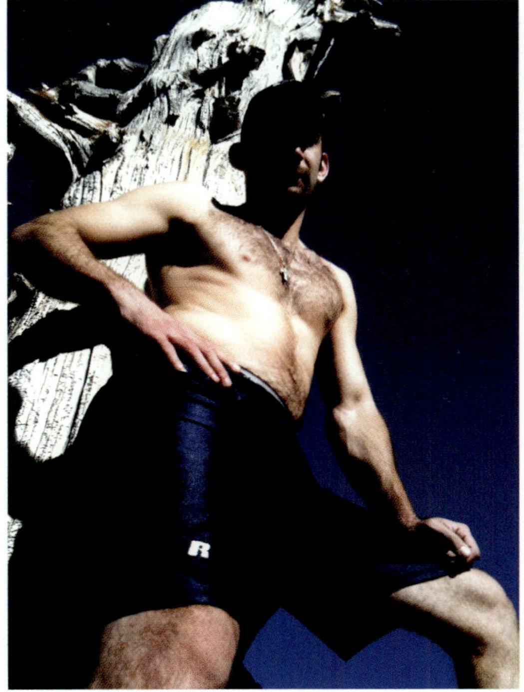
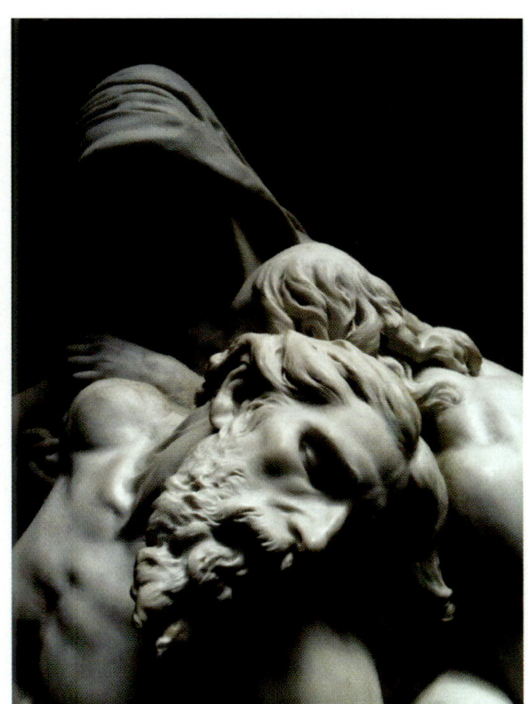

FOOD

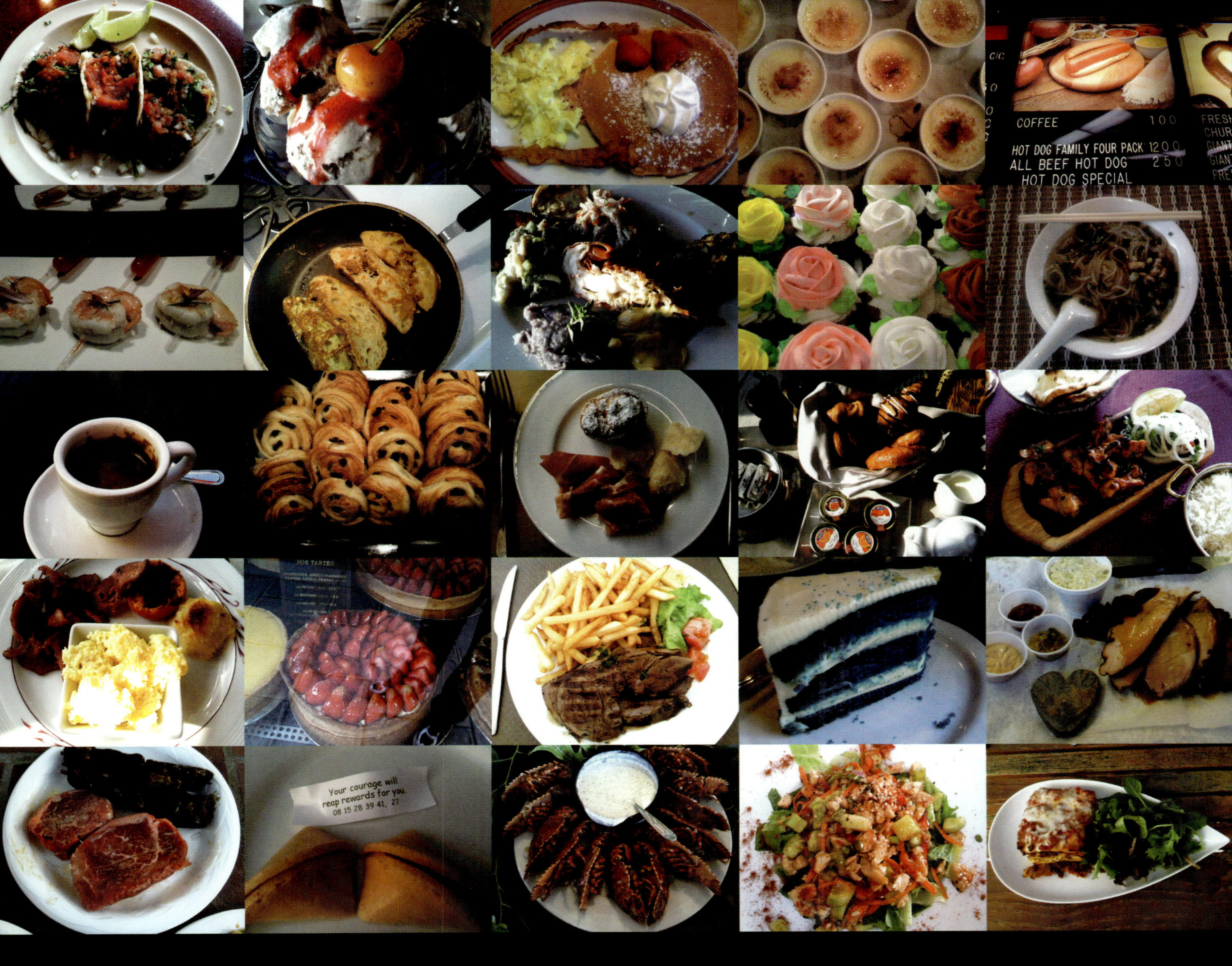
COFFEE 100
HOT DOG FAMILY FOUR PACK 1200
ALL BEEF HOT DOG 250
HOT DOG SPECIAL
FRESH
CHUR
GIANT
GIANT
FRE
Your courage will
reap rewards for you.
08 15 28 39 41 27

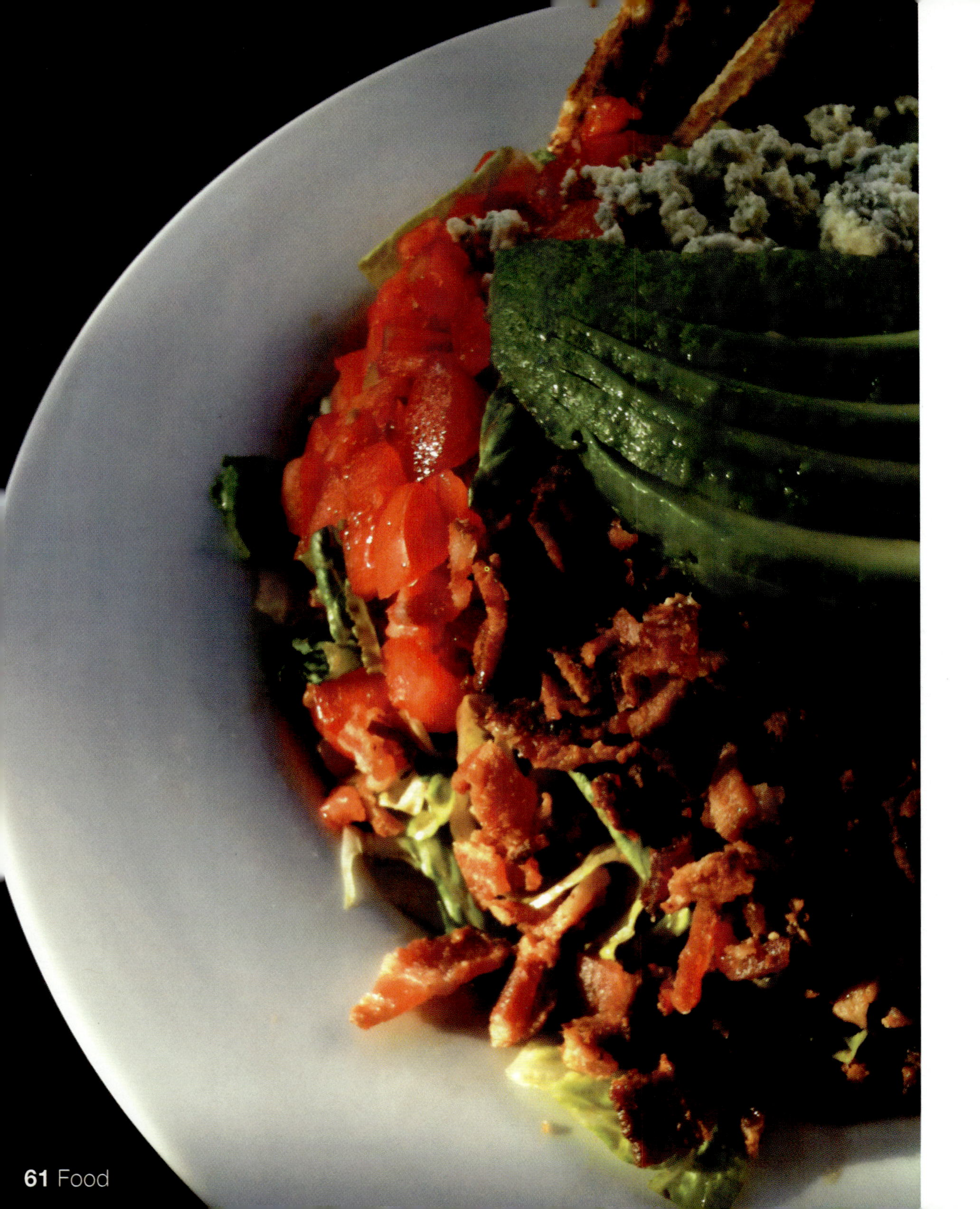

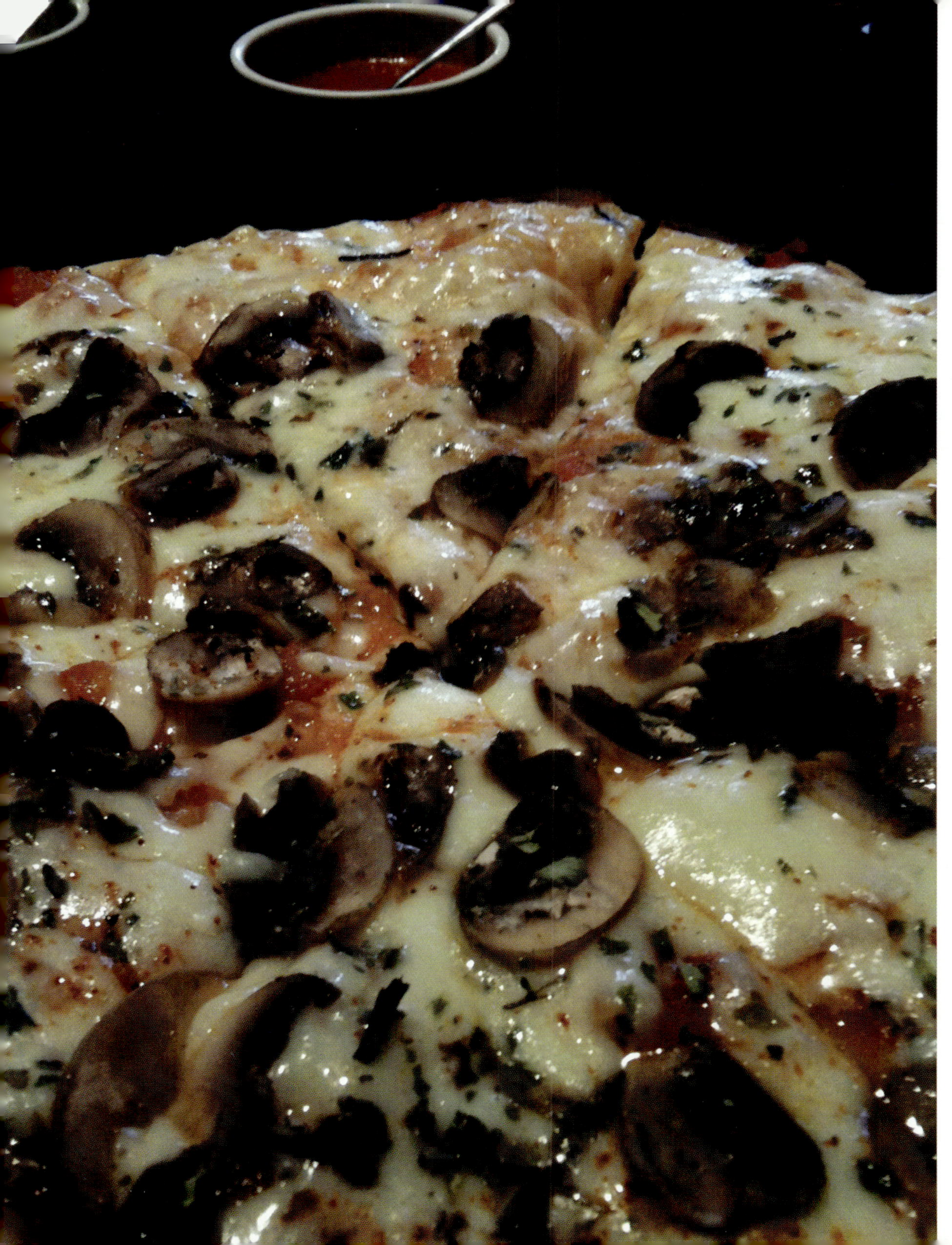

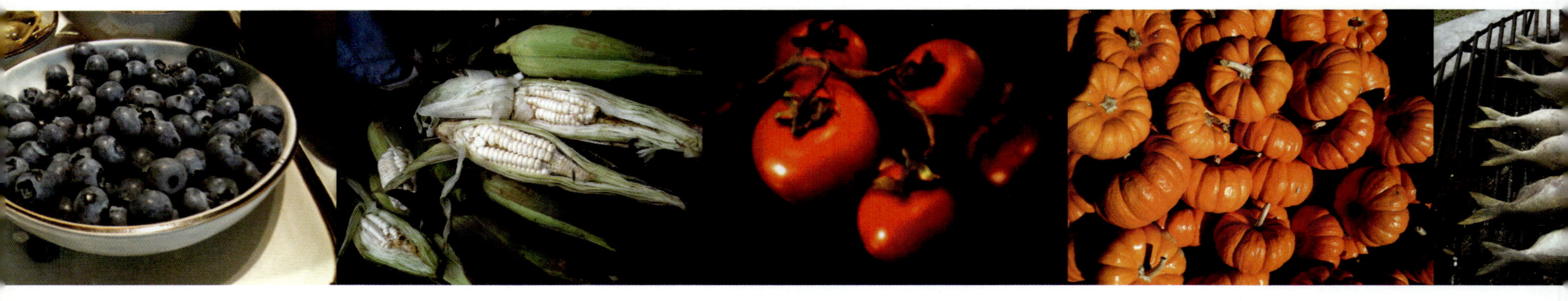

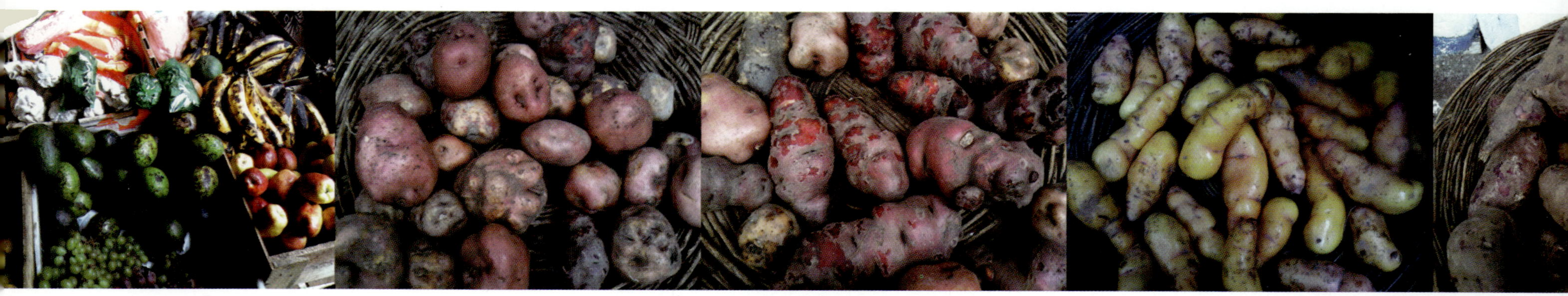

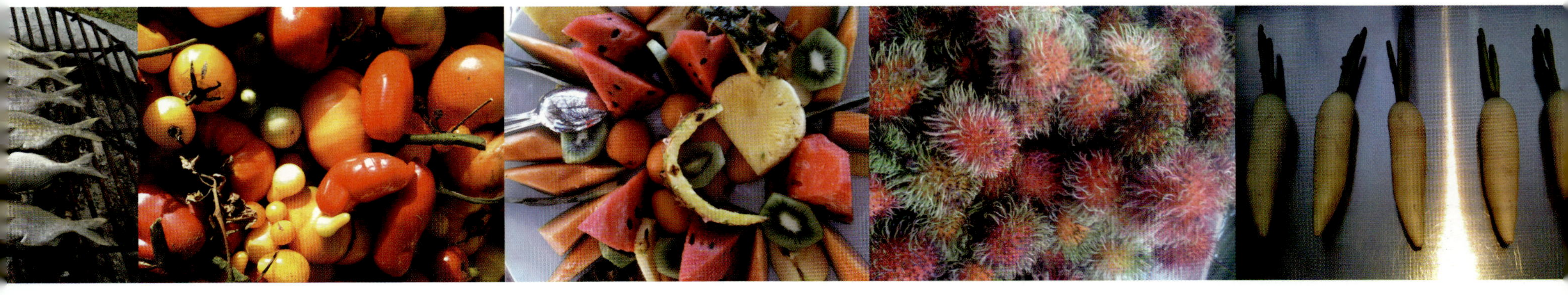

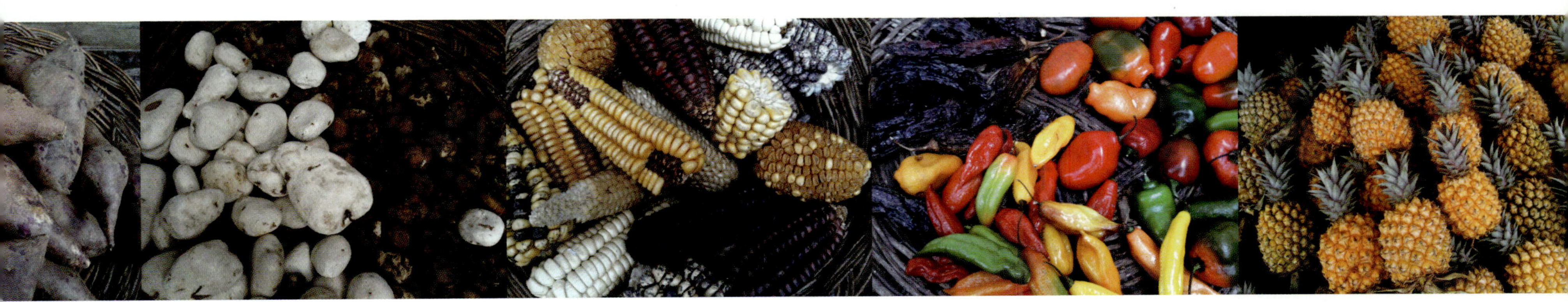

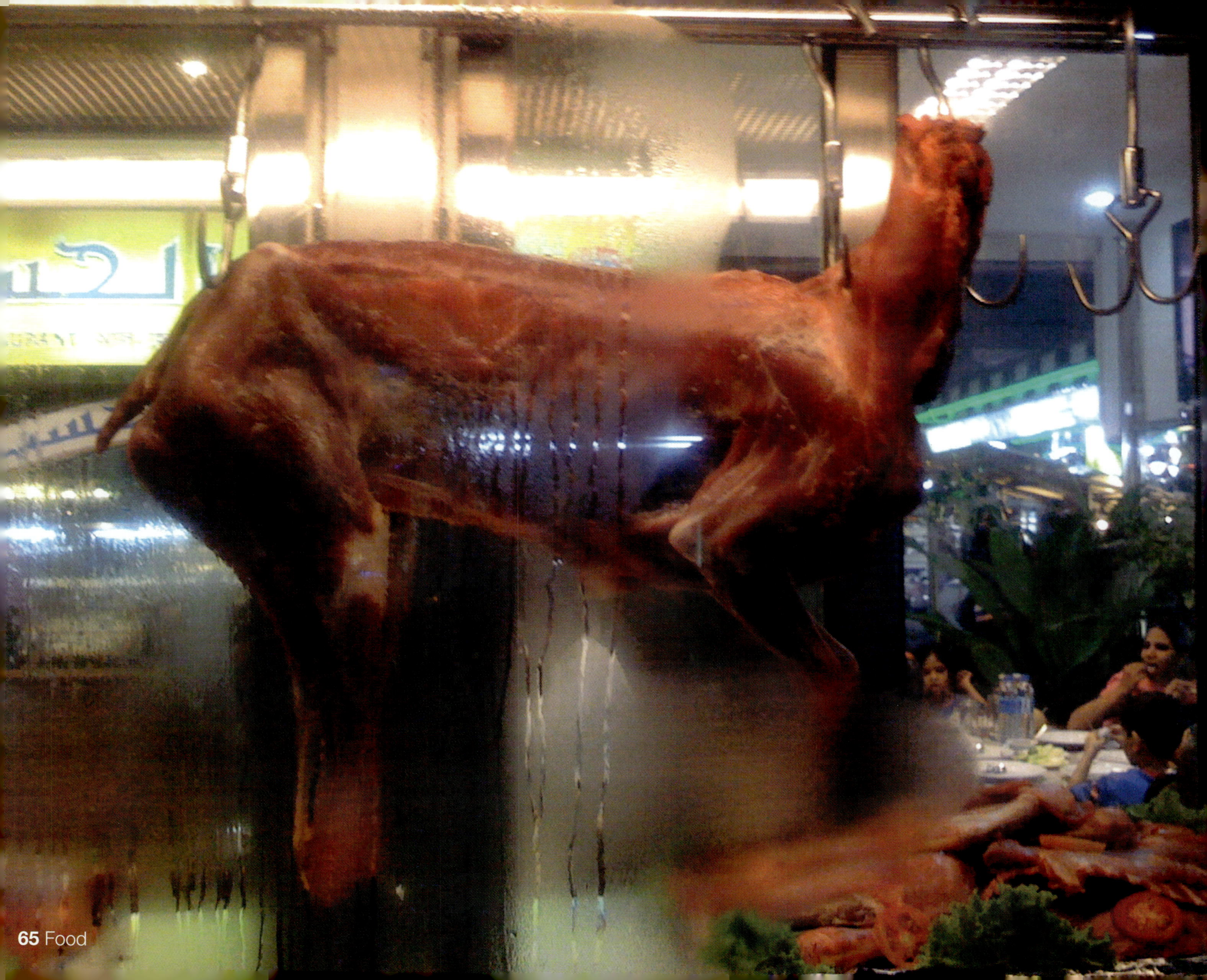

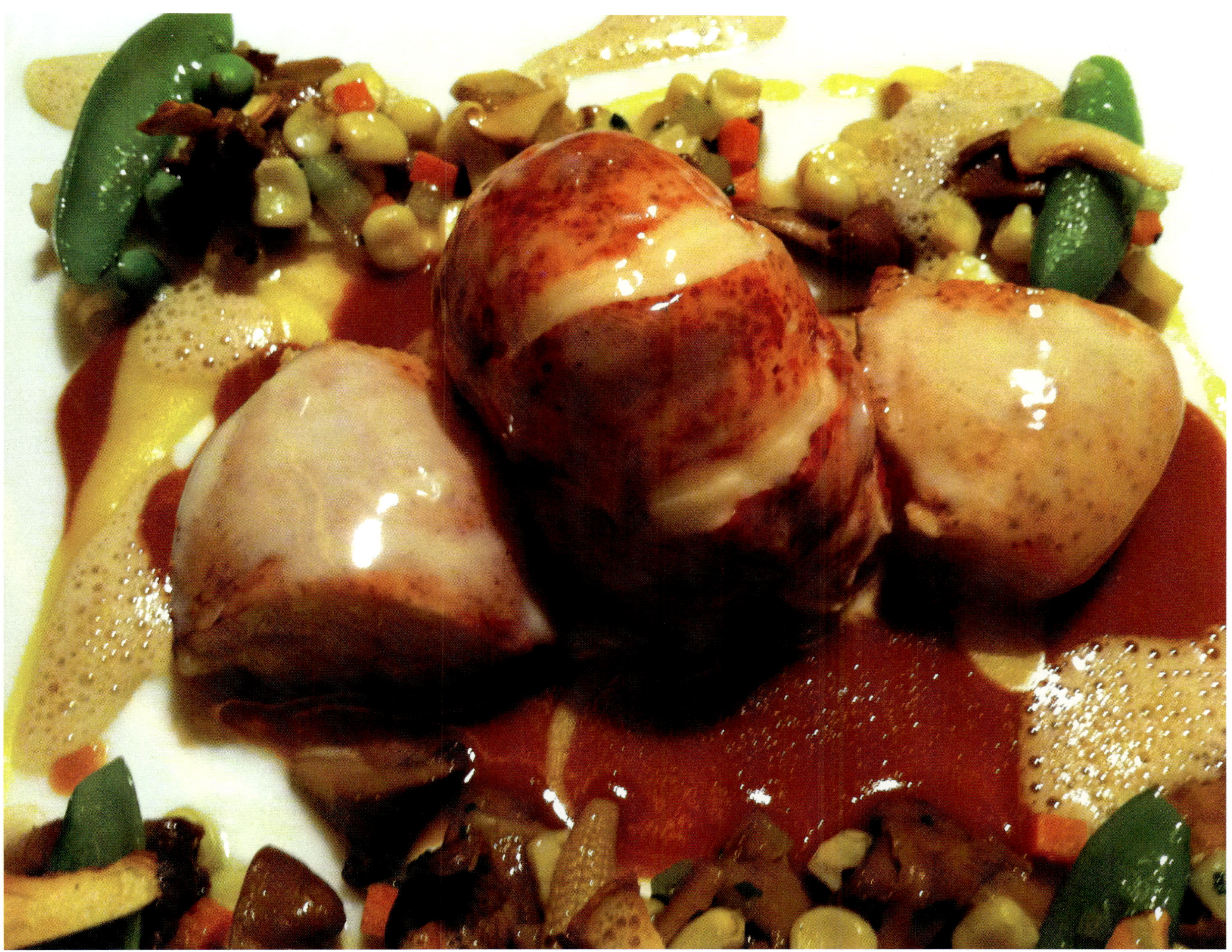

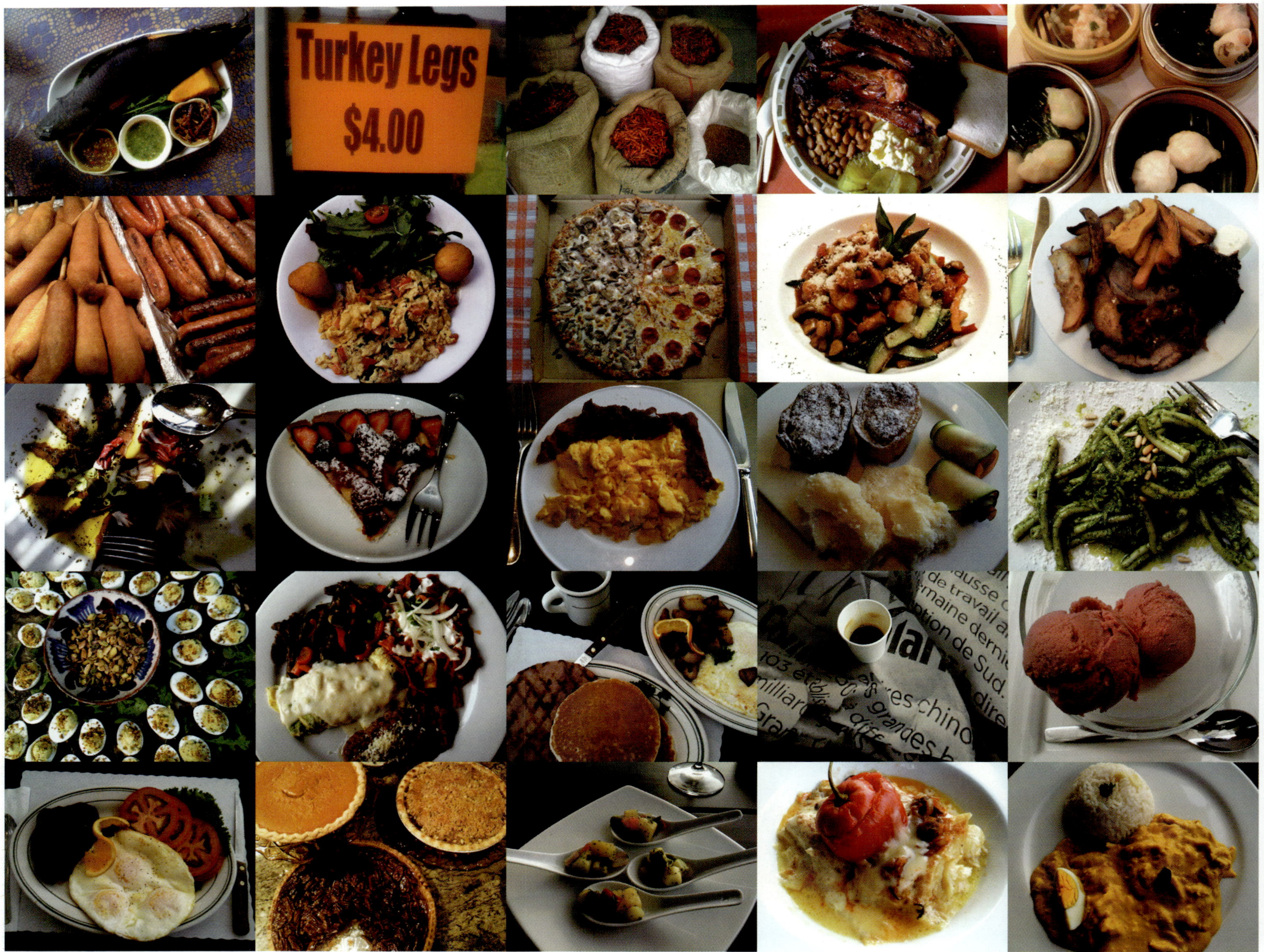

Turkey Legs
$4.00

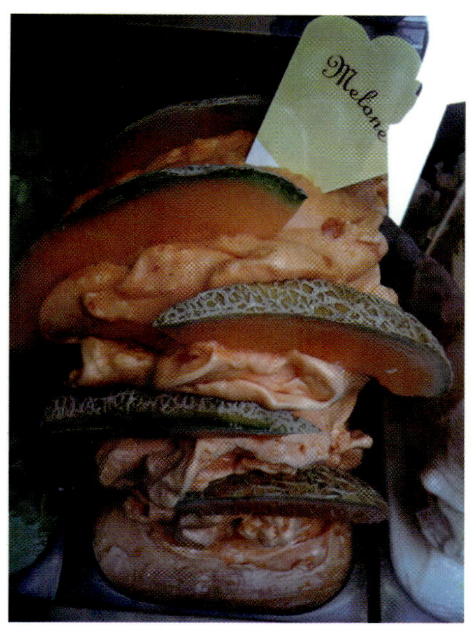 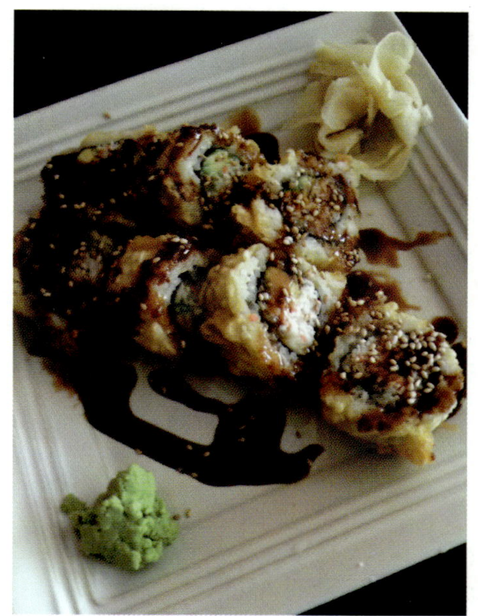 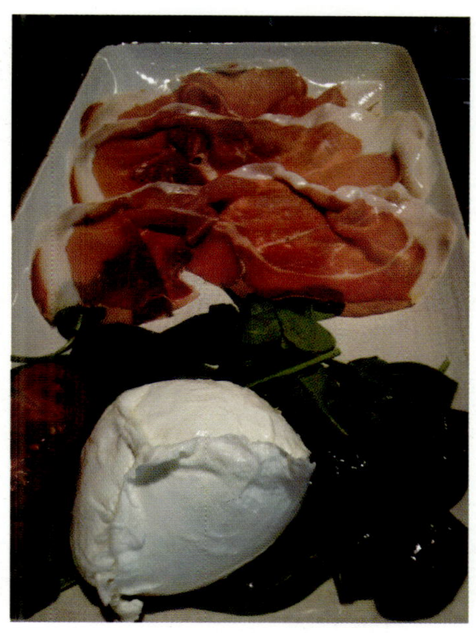

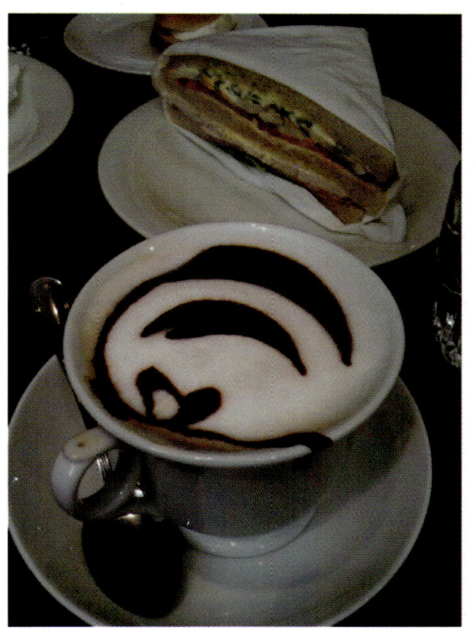 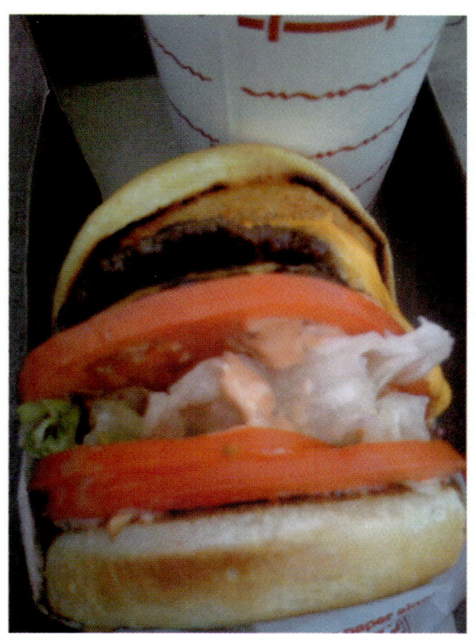 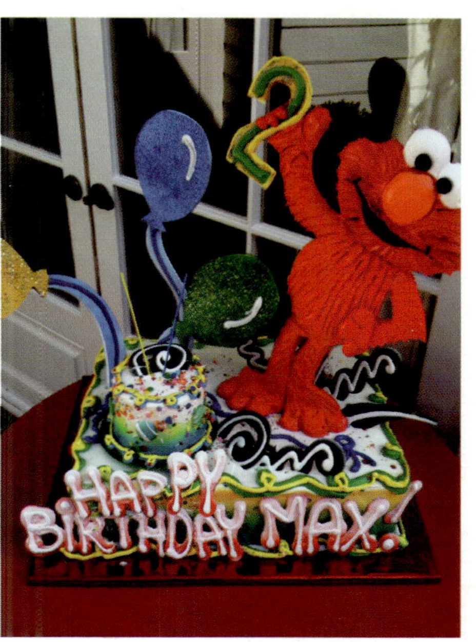

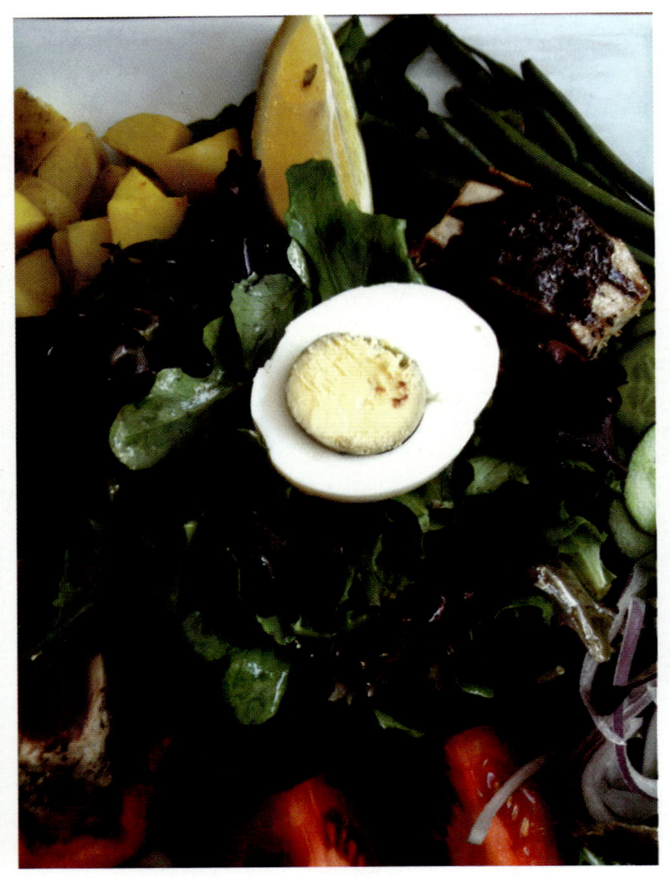
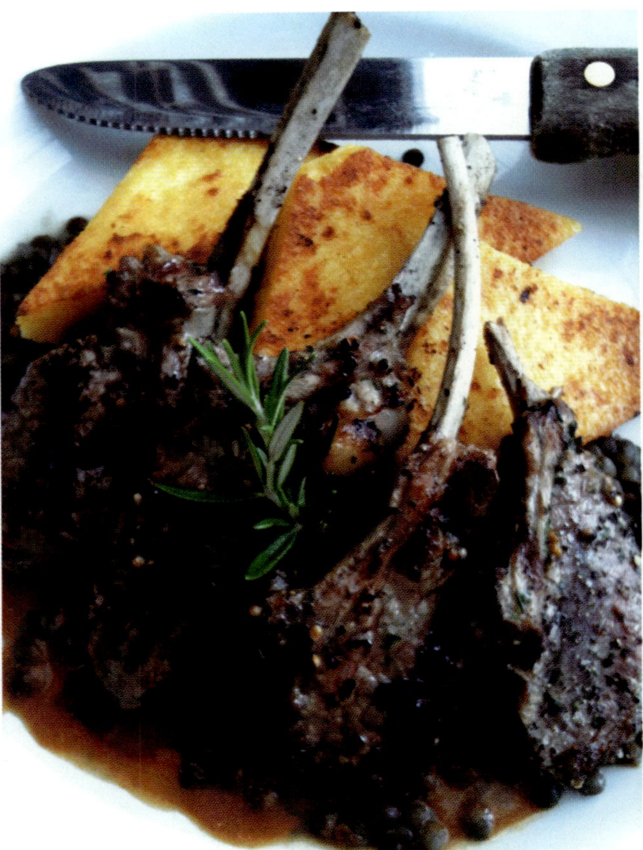
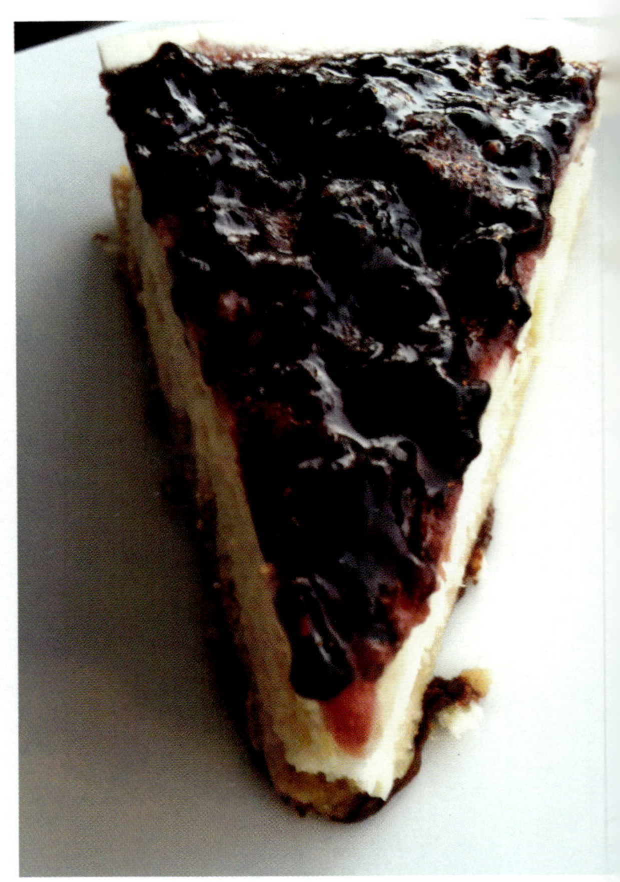

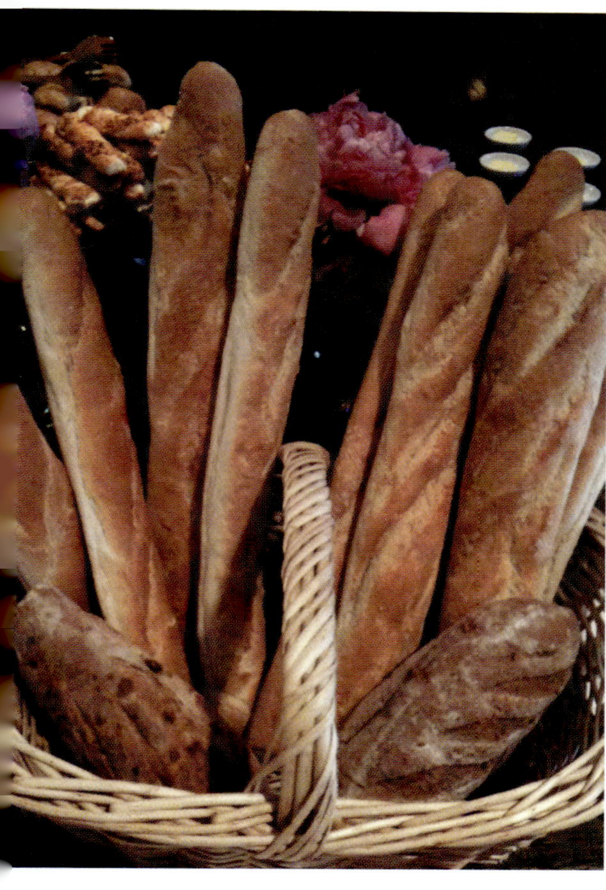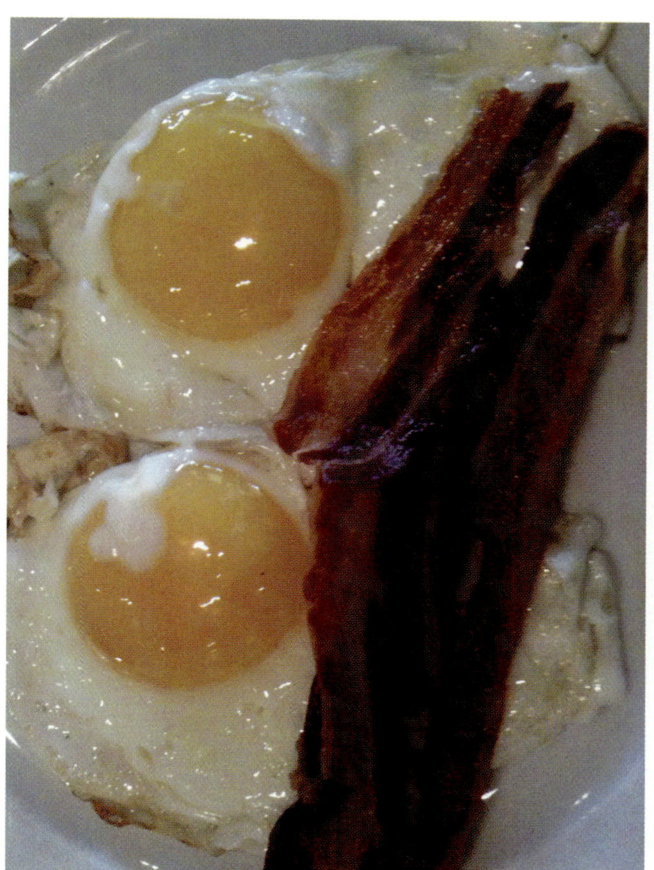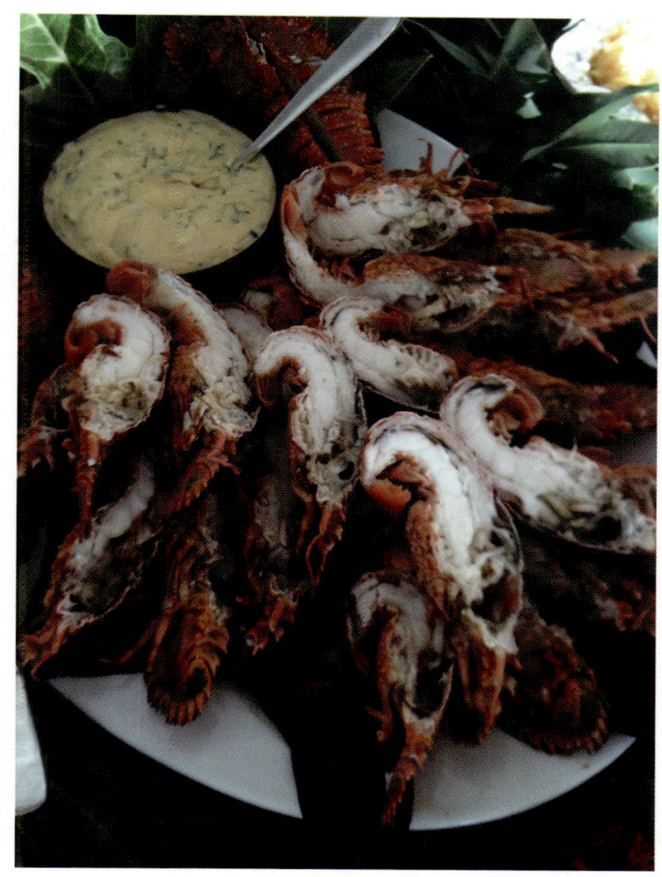

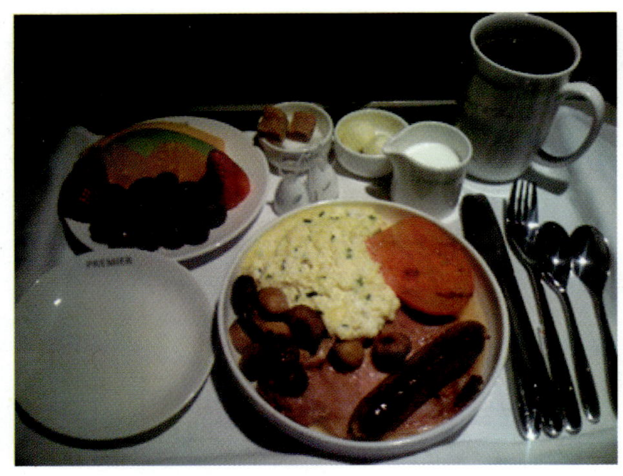
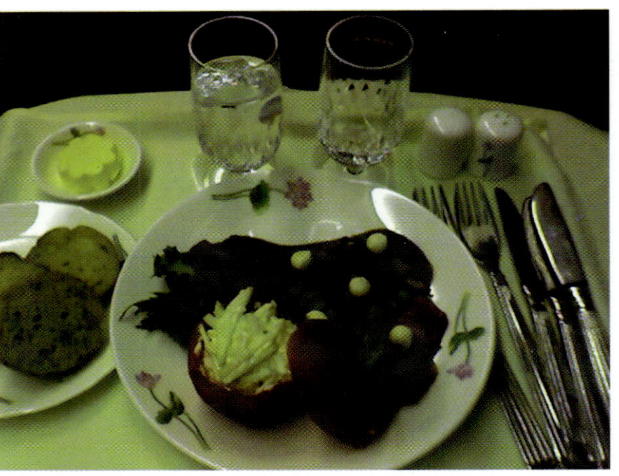
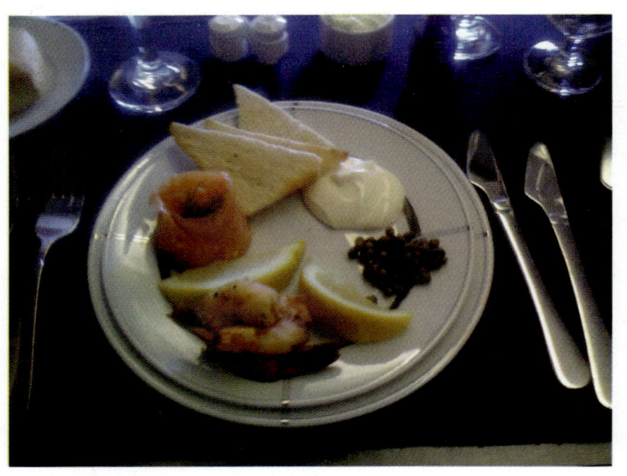
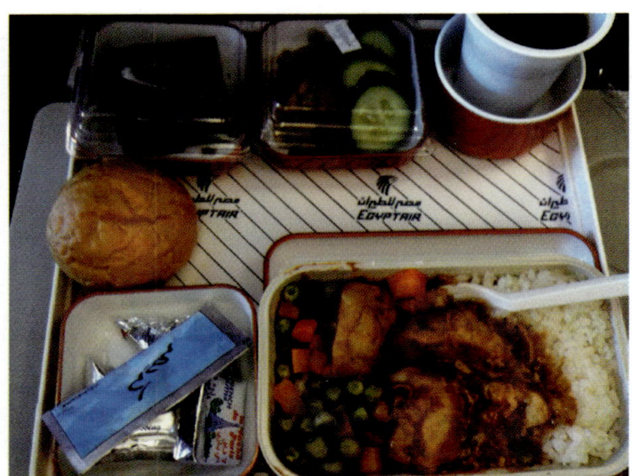
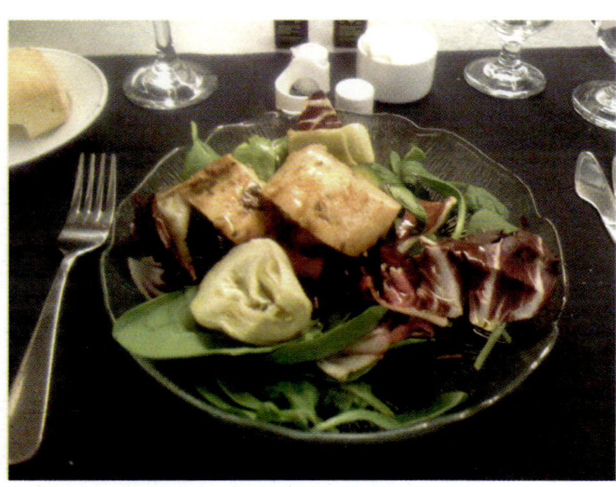
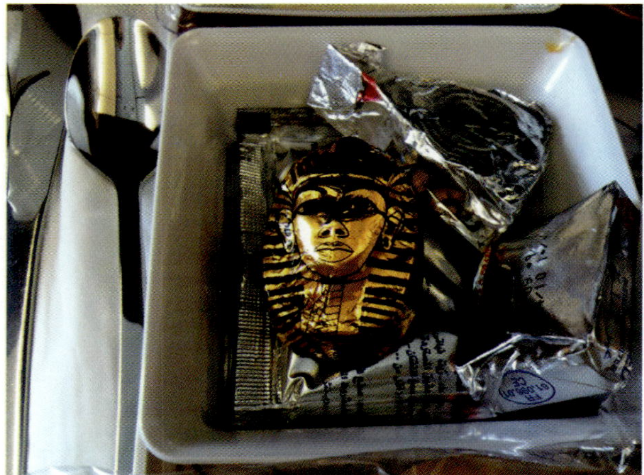

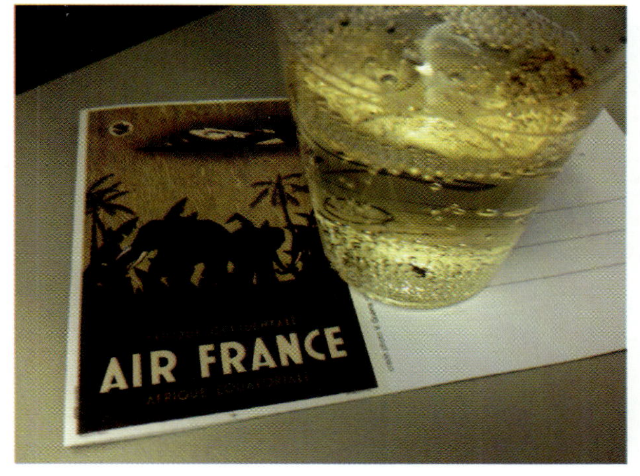
AIR FRANCE

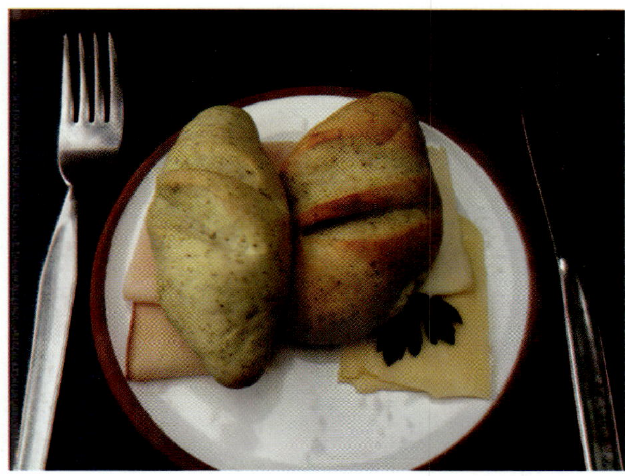

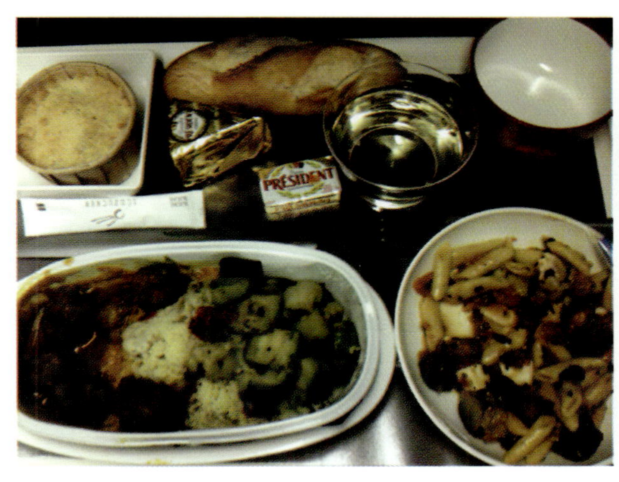
PRÉSIDENT

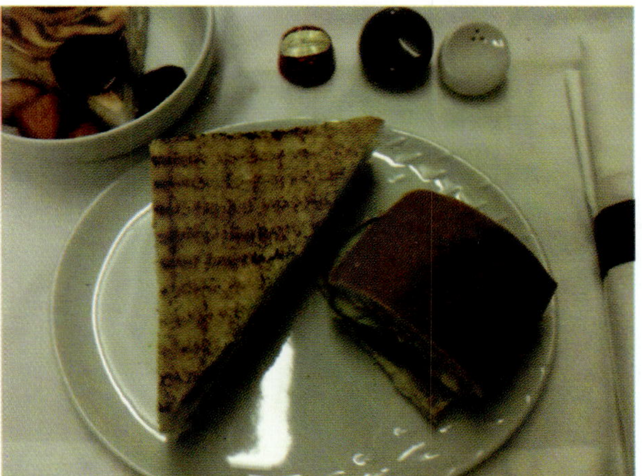

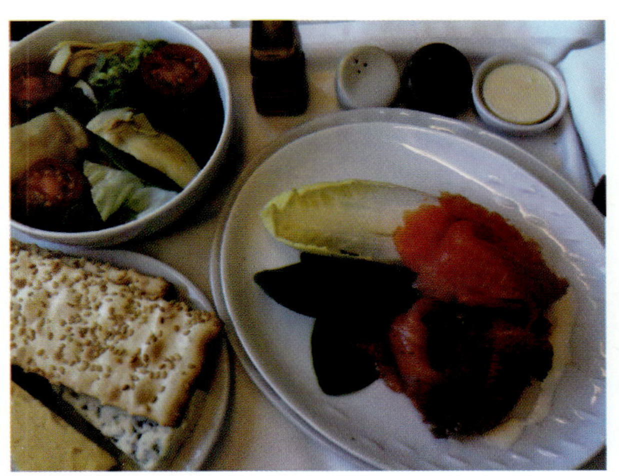

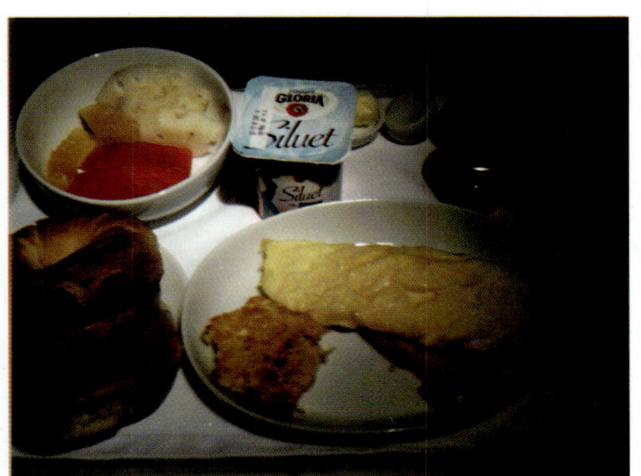
GLORIA
Siluet

LIPS

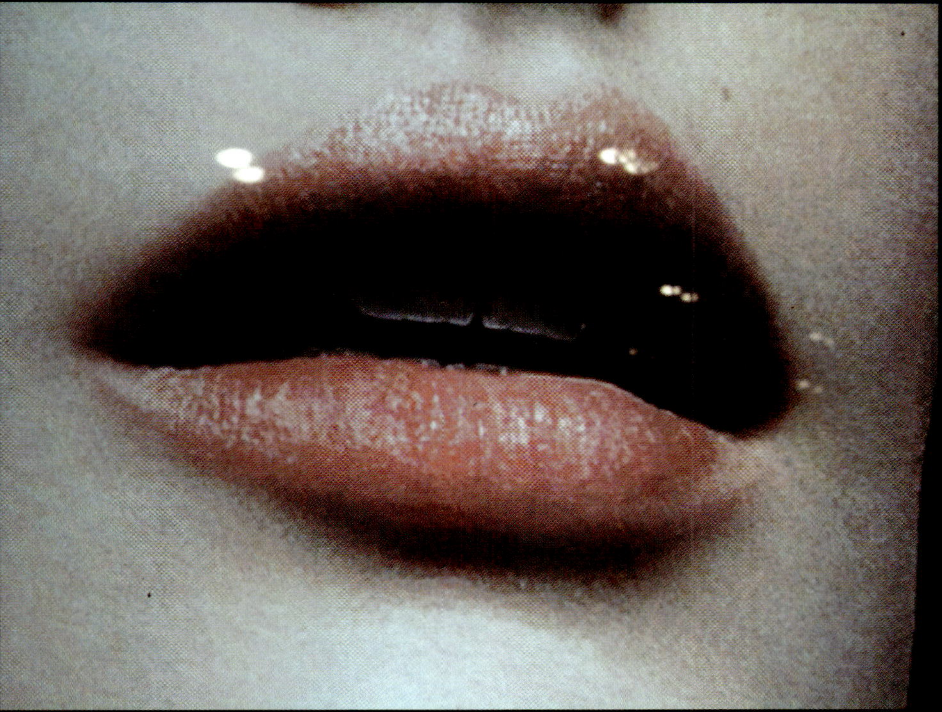

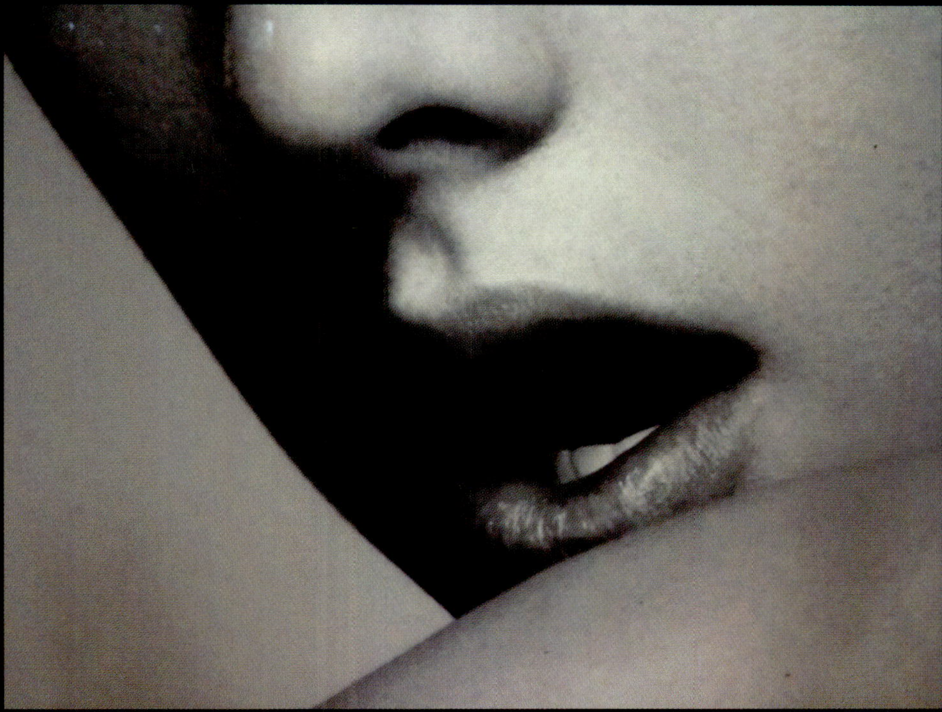

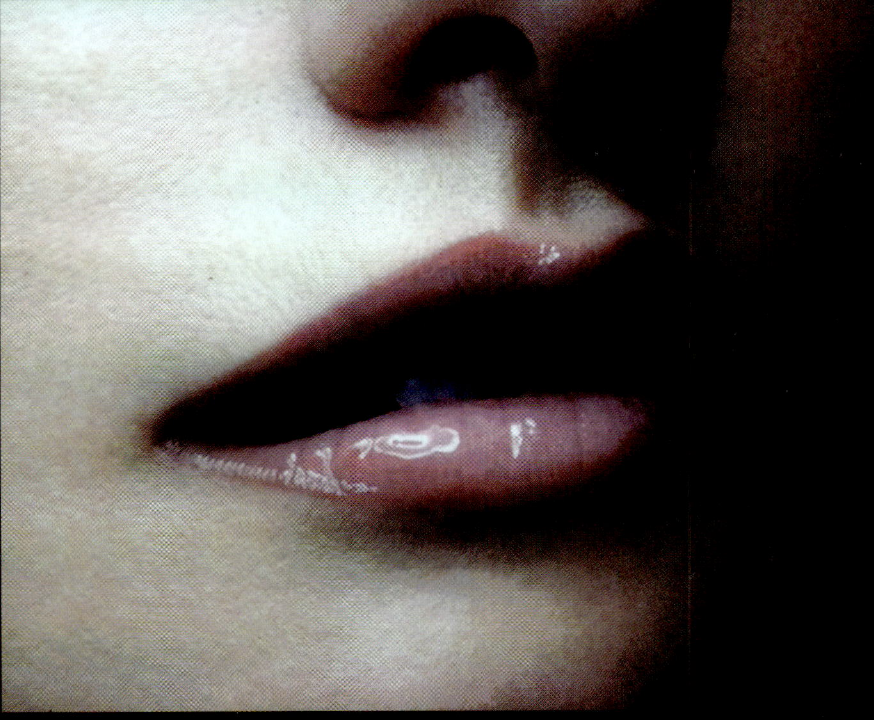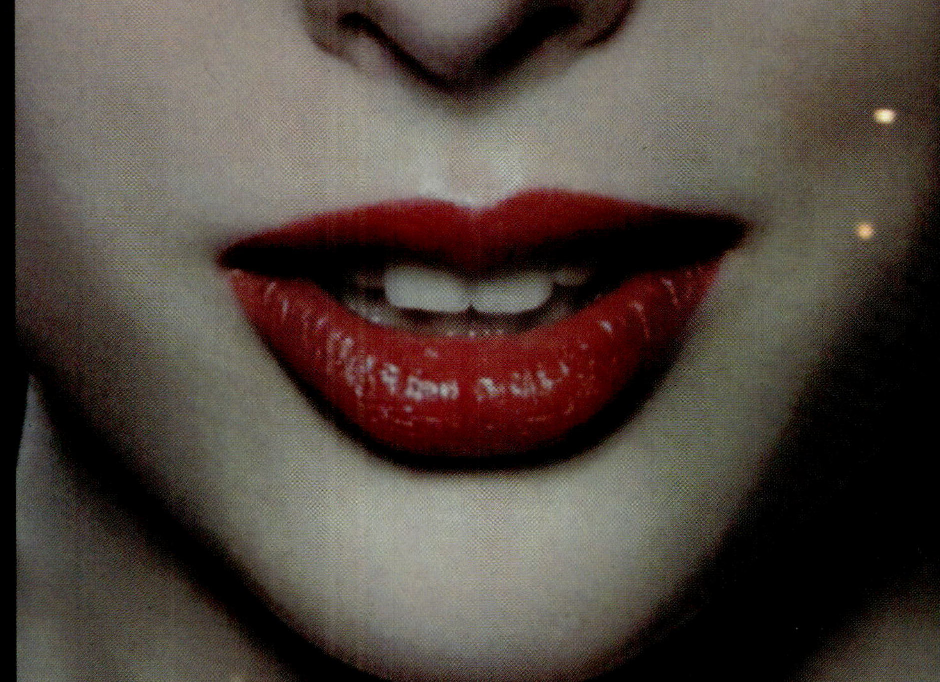

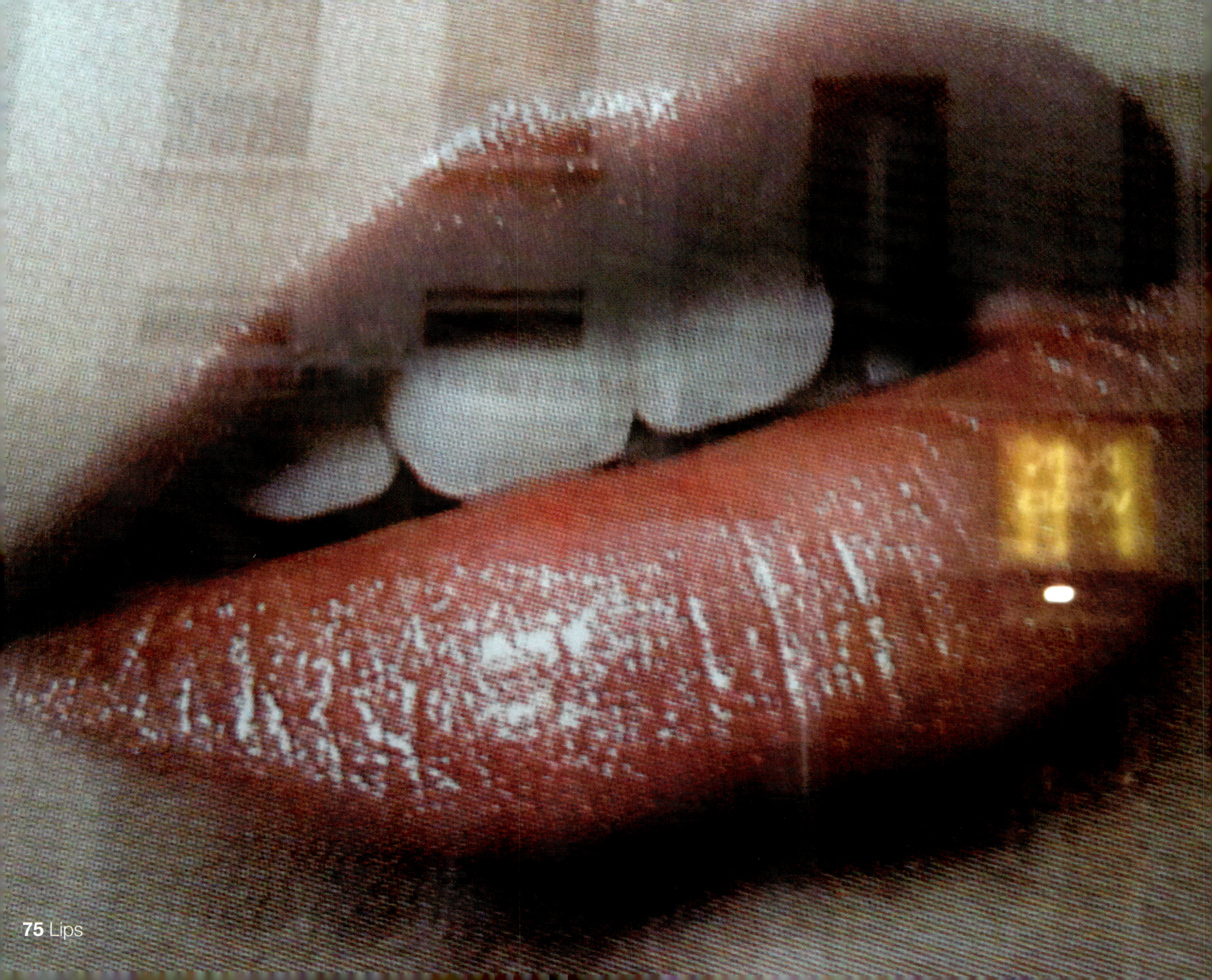

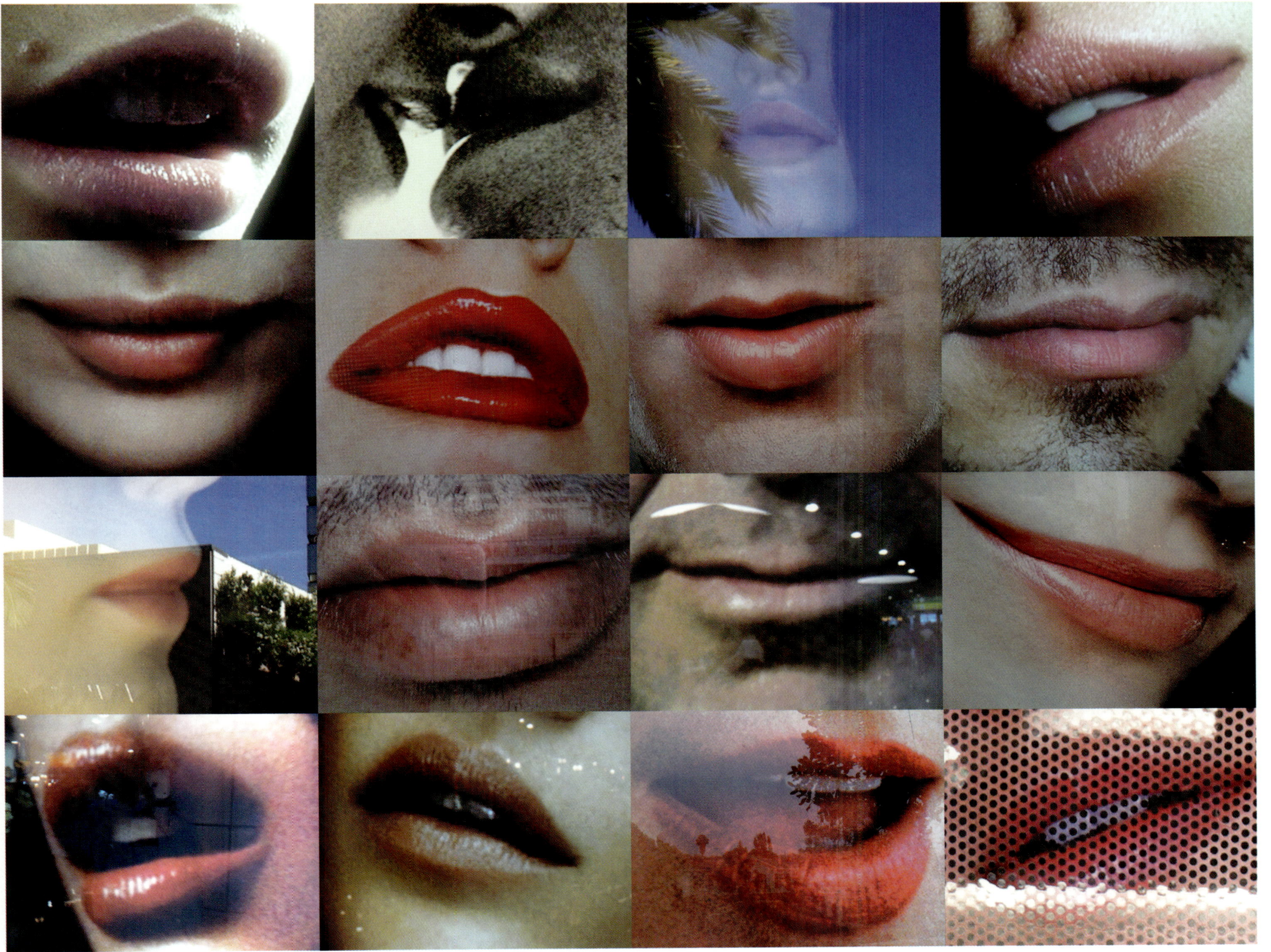

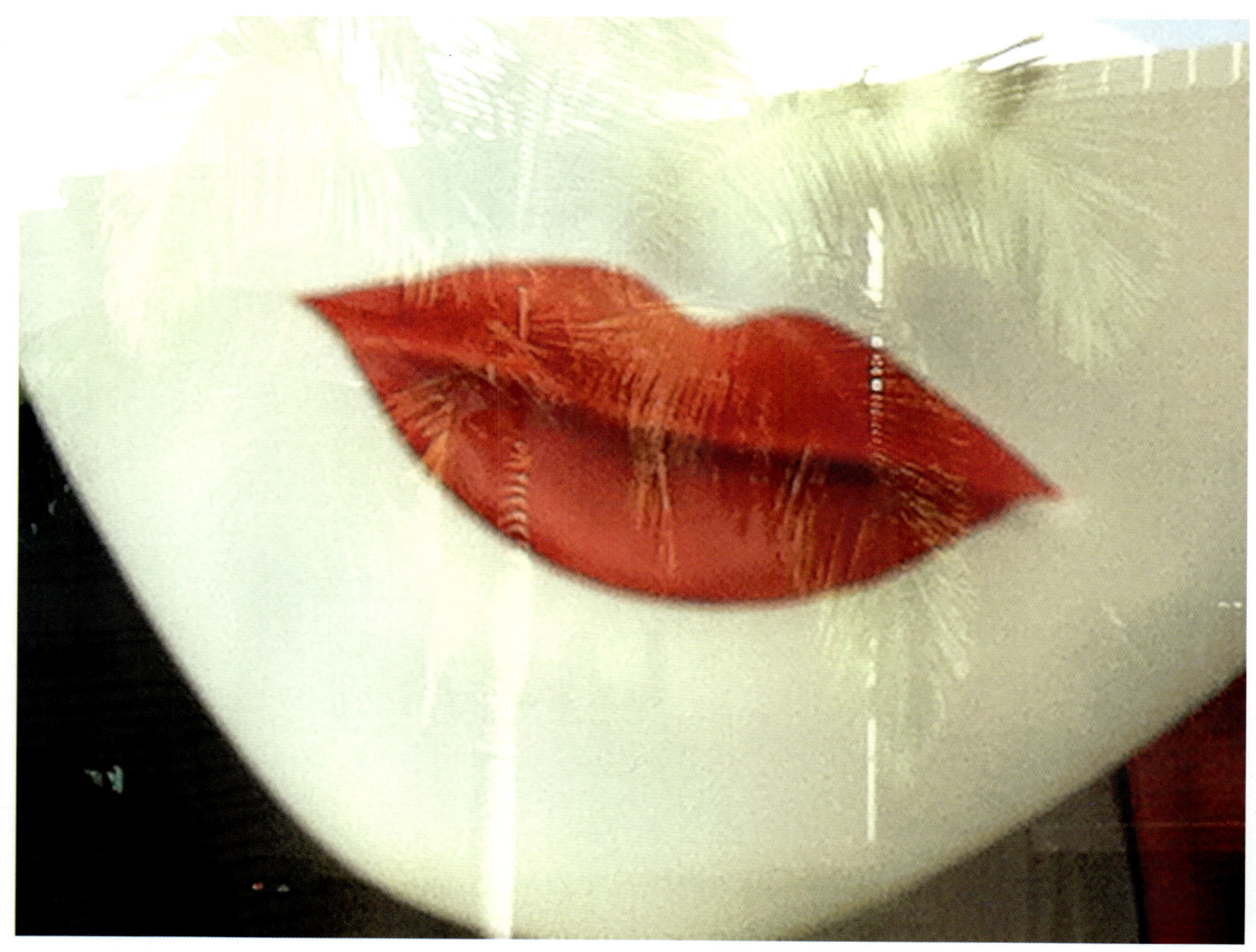

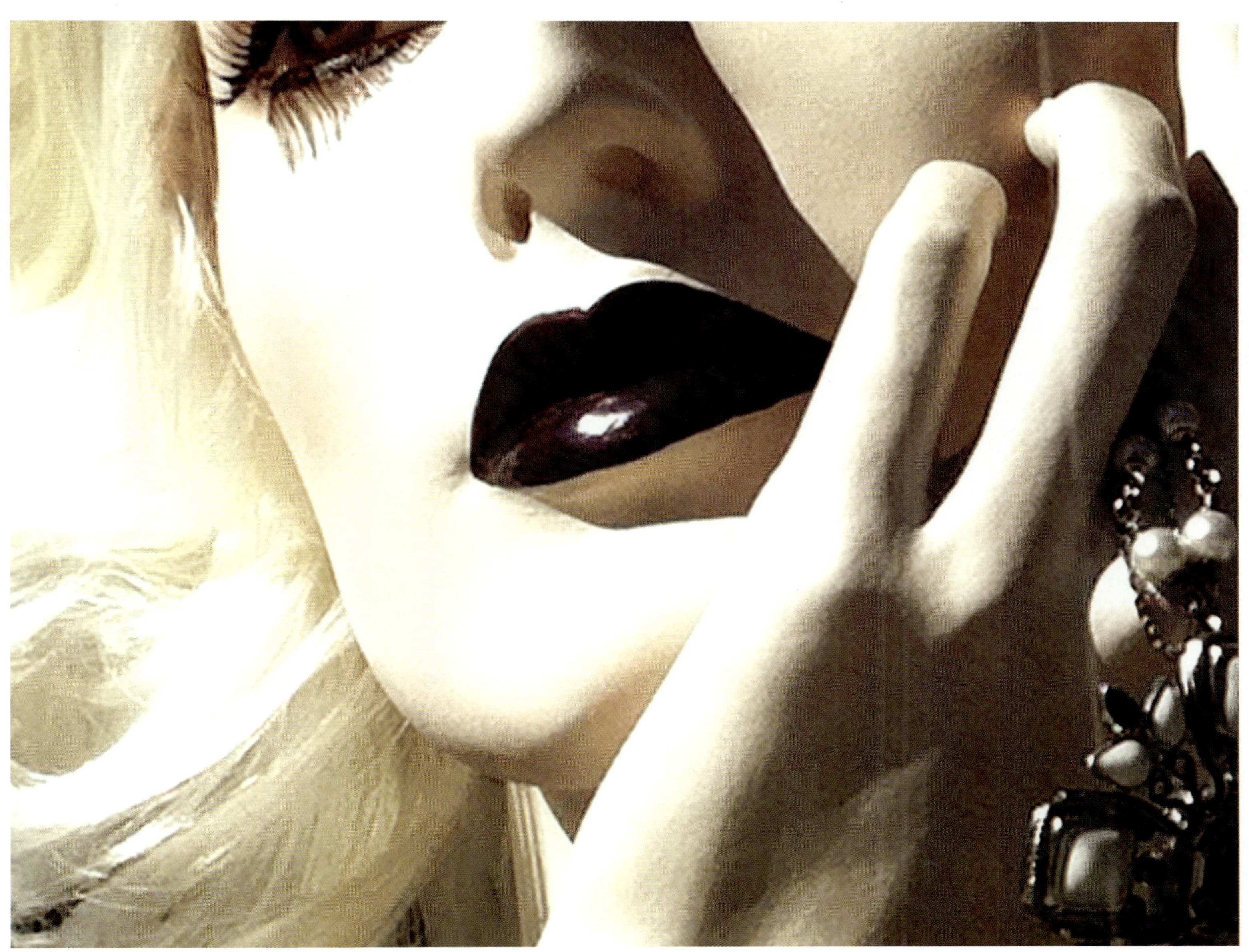

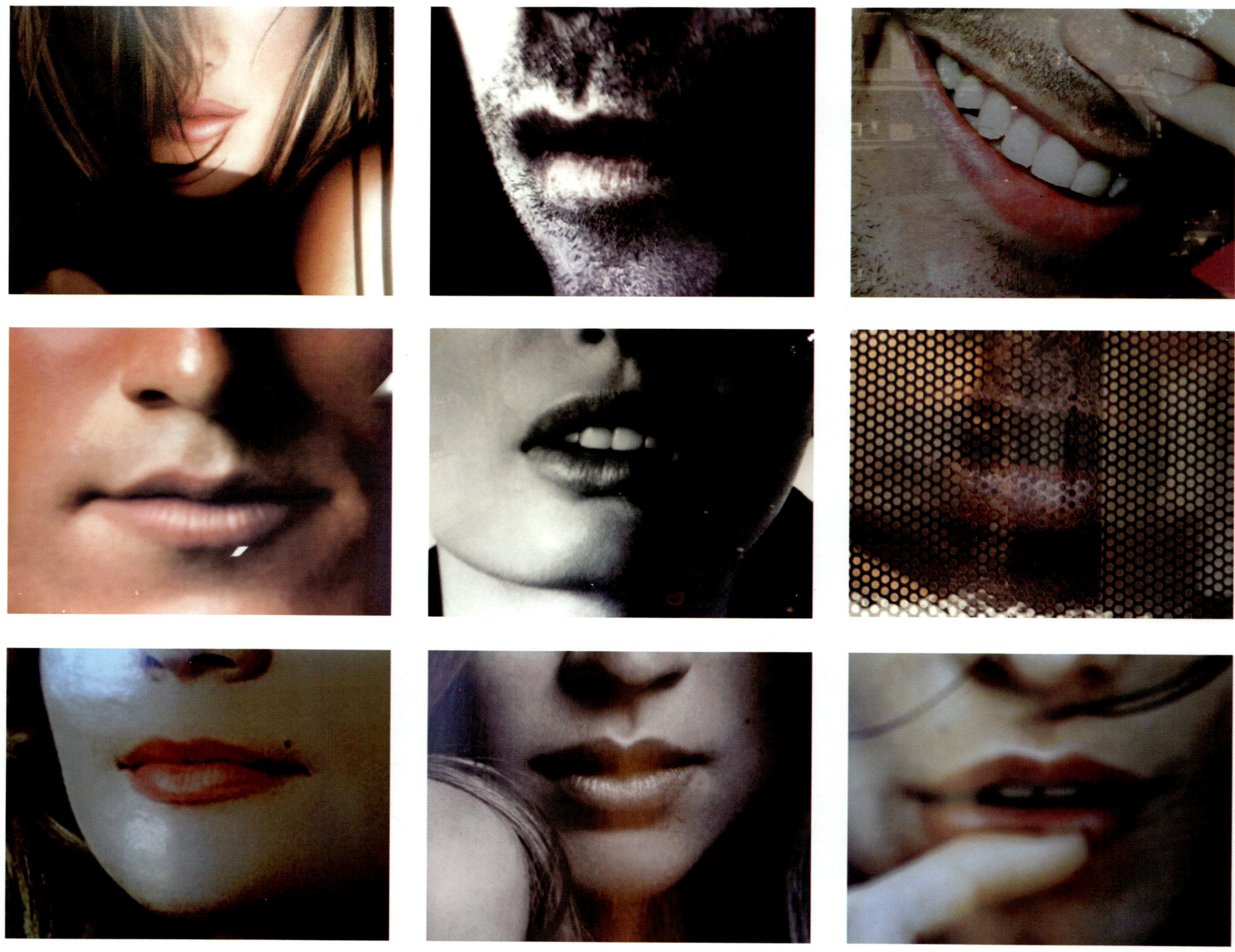

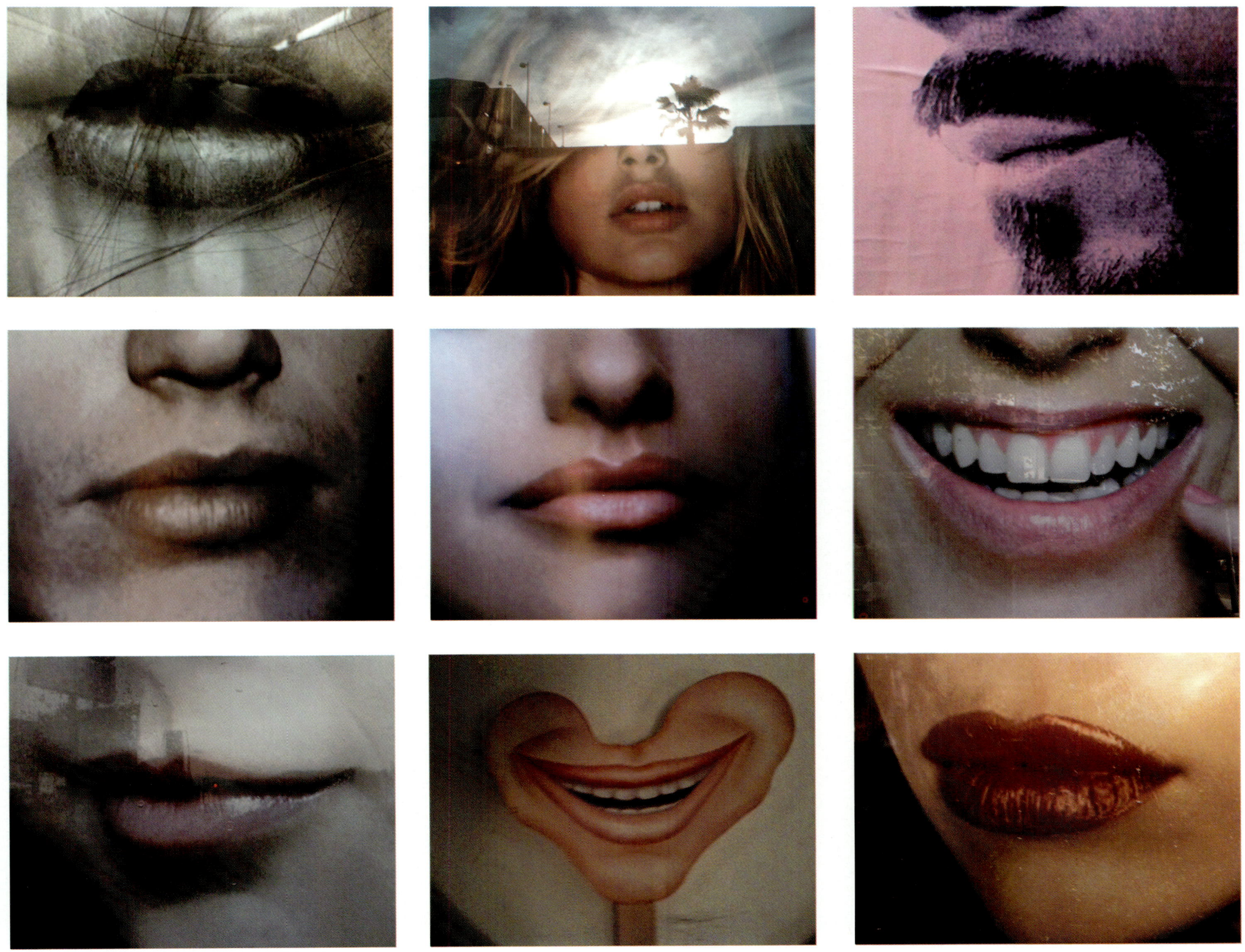

ME (PHOTO DIARY)

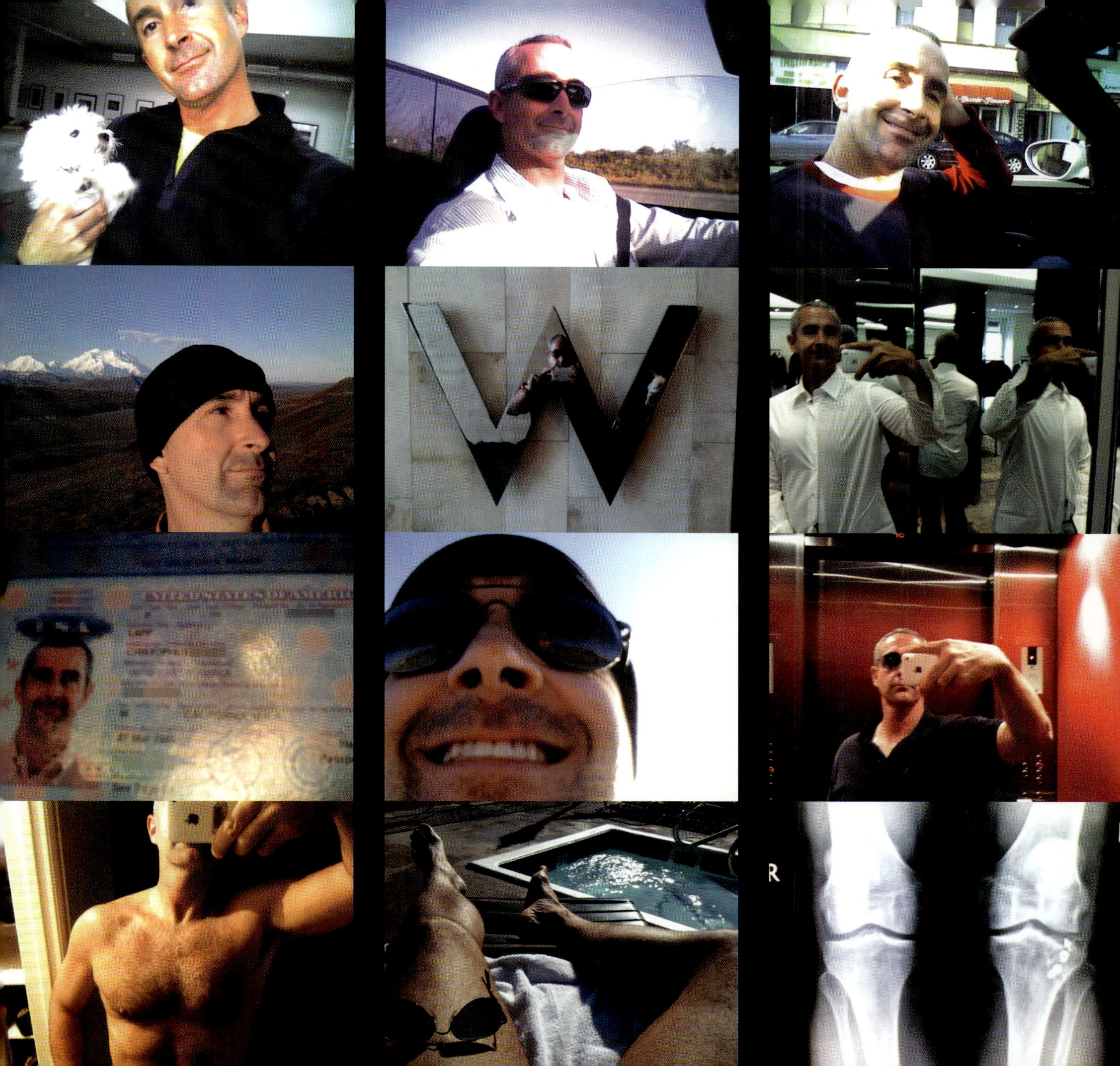

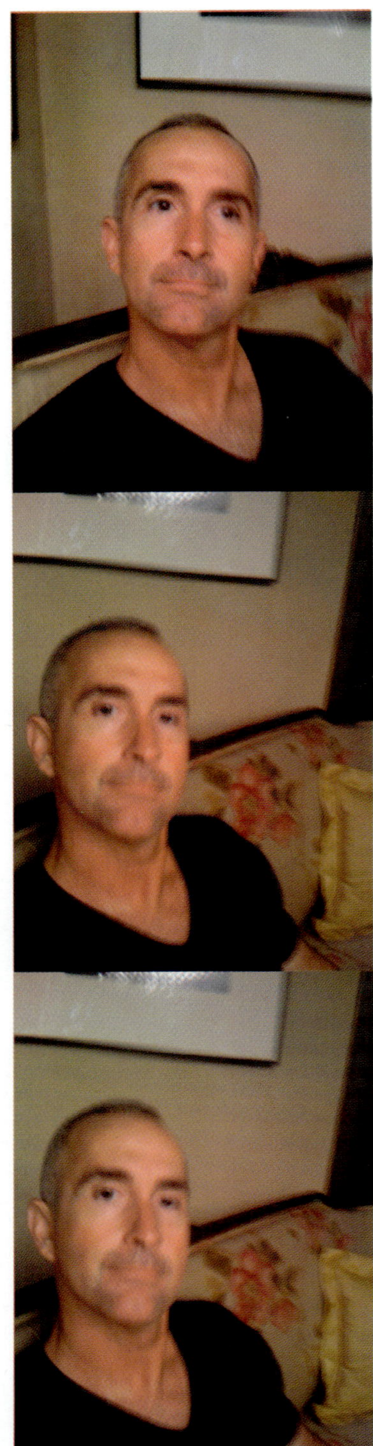
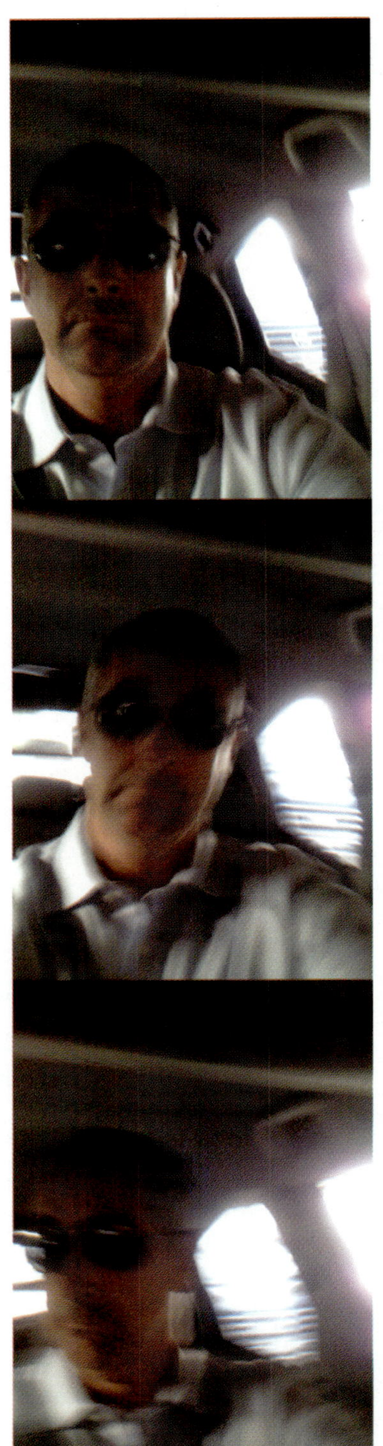
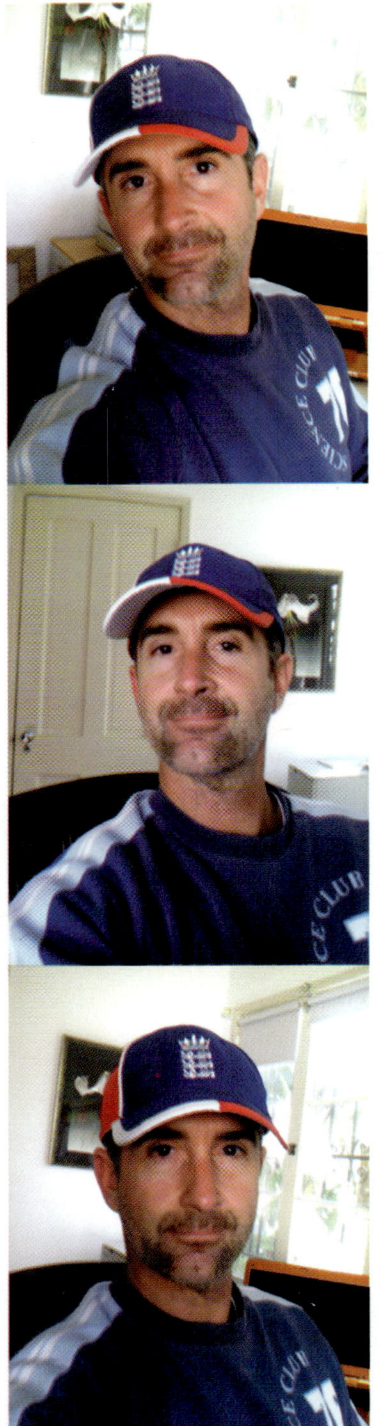
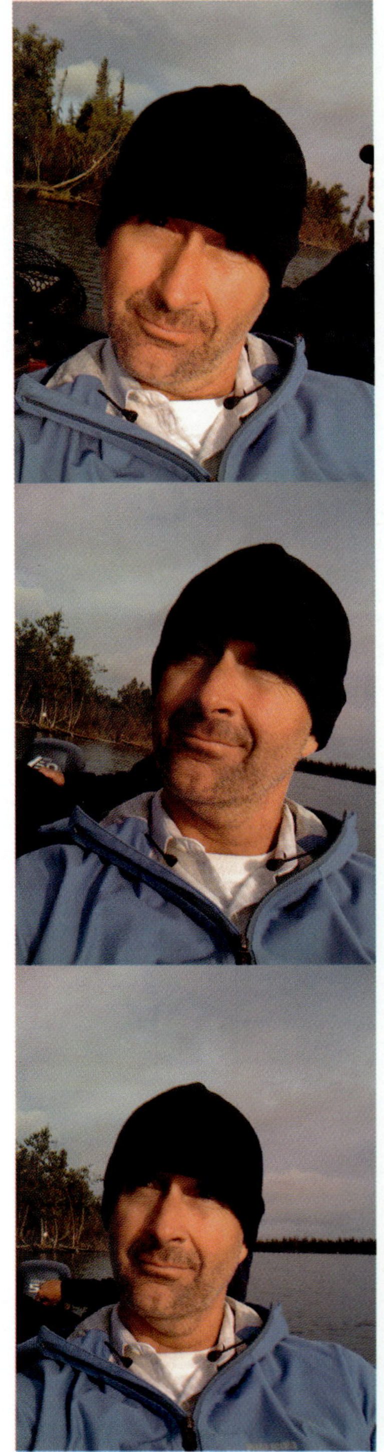

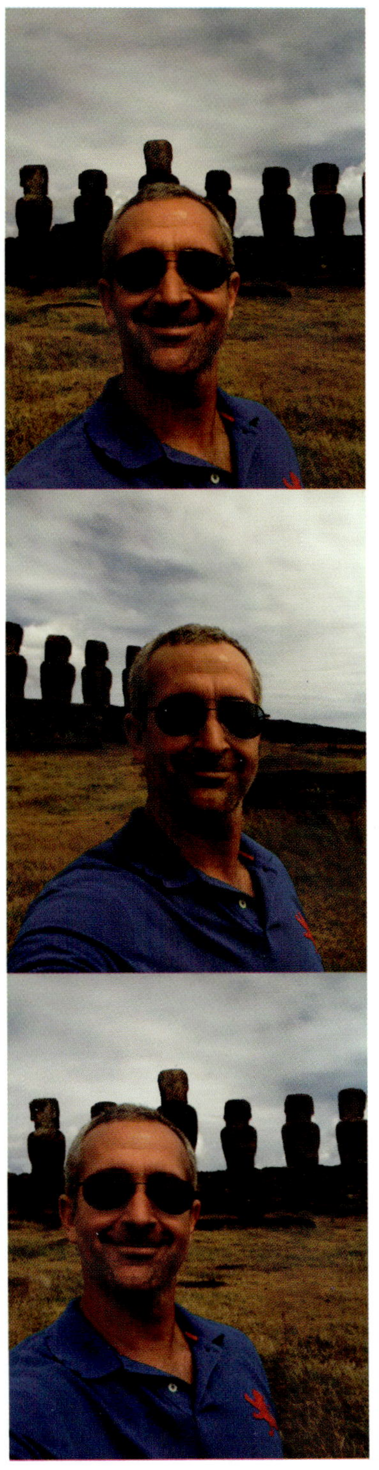

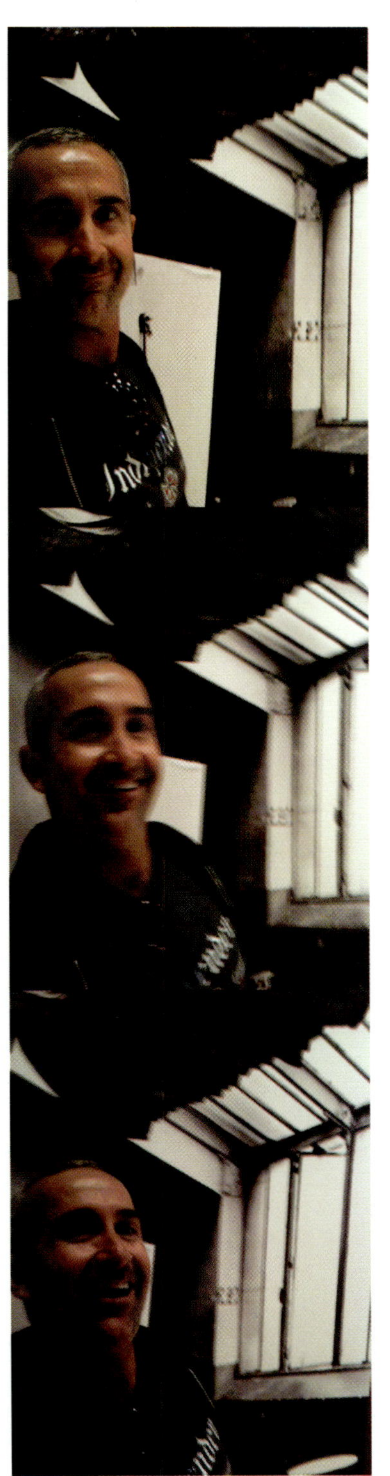

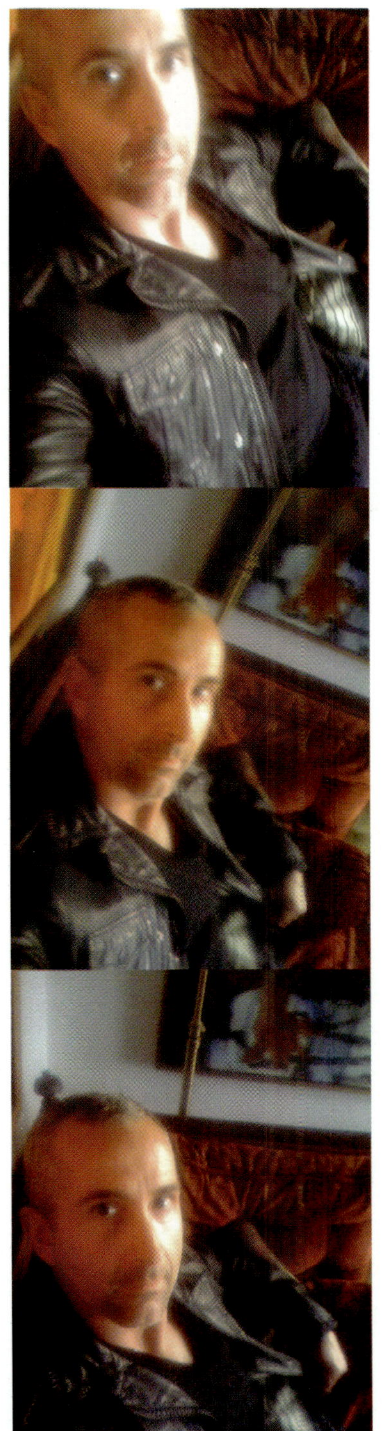

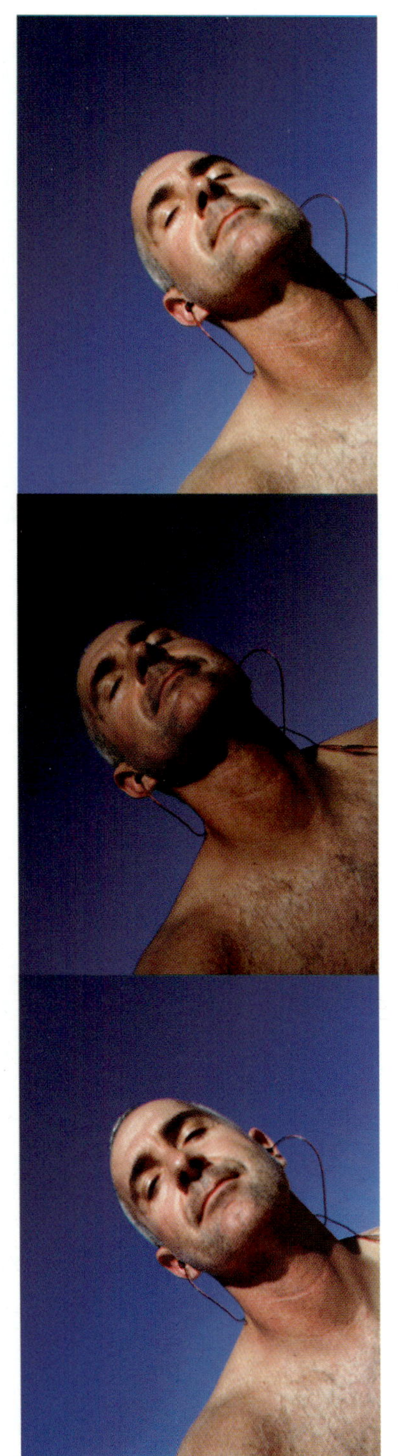

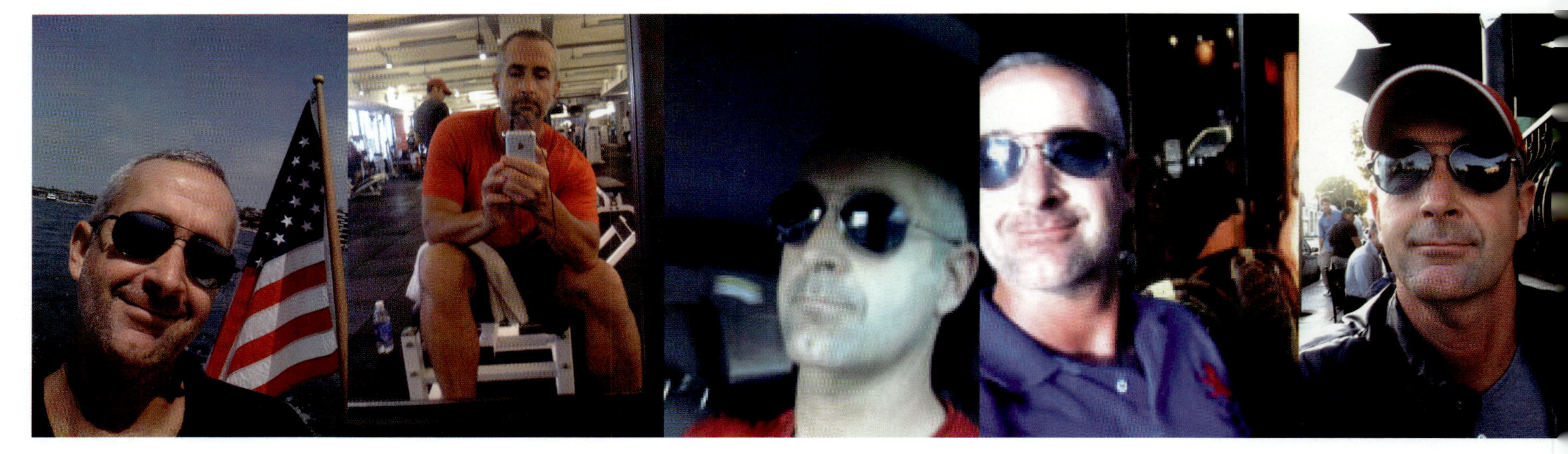

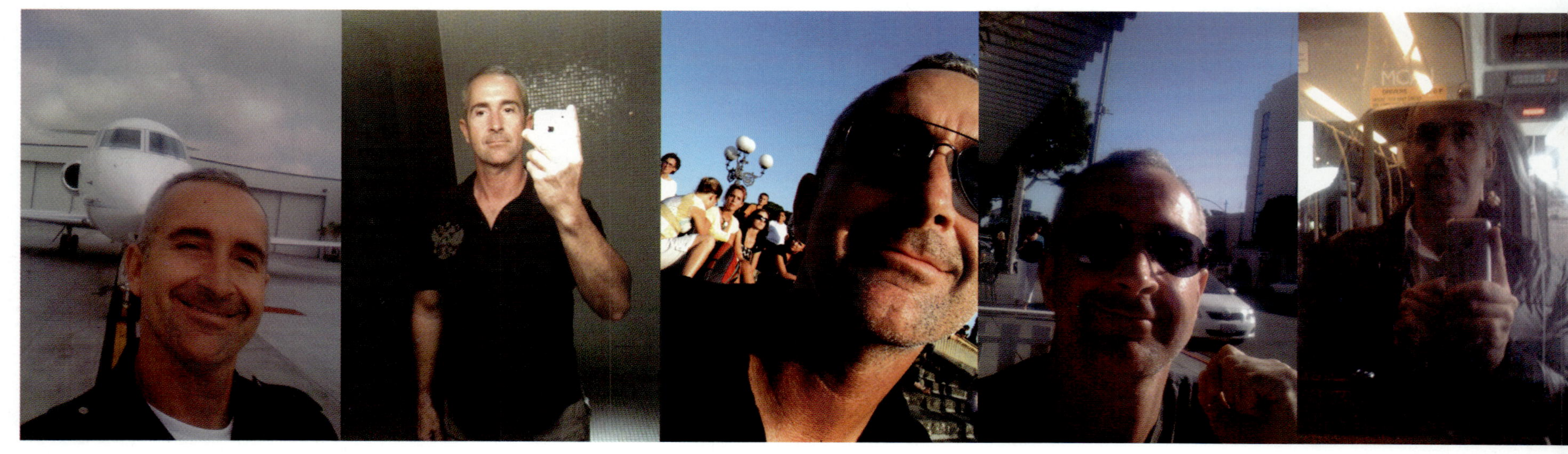

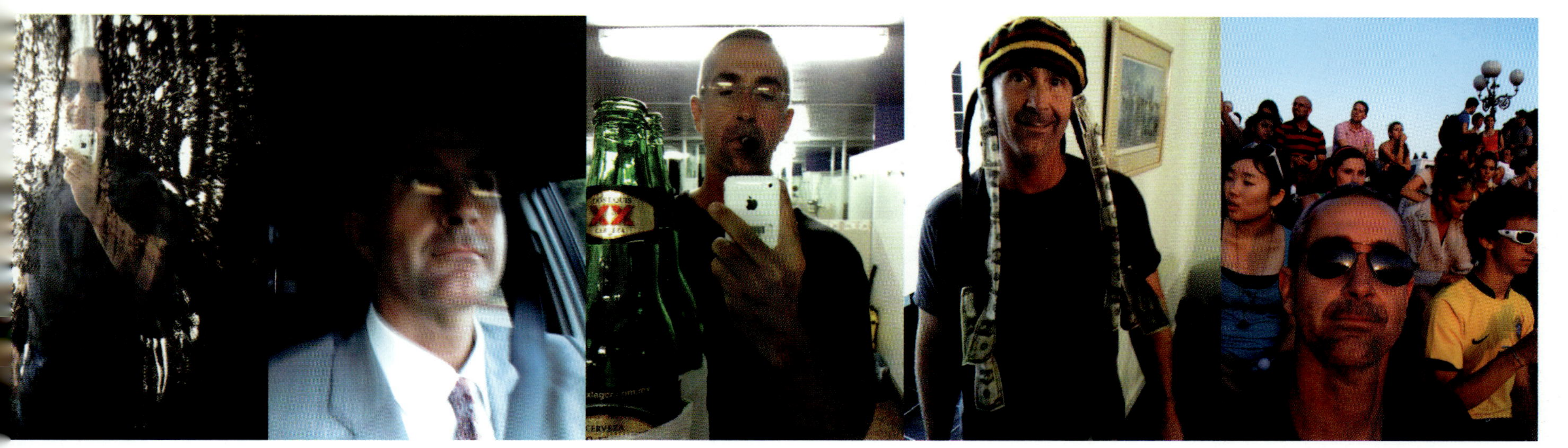

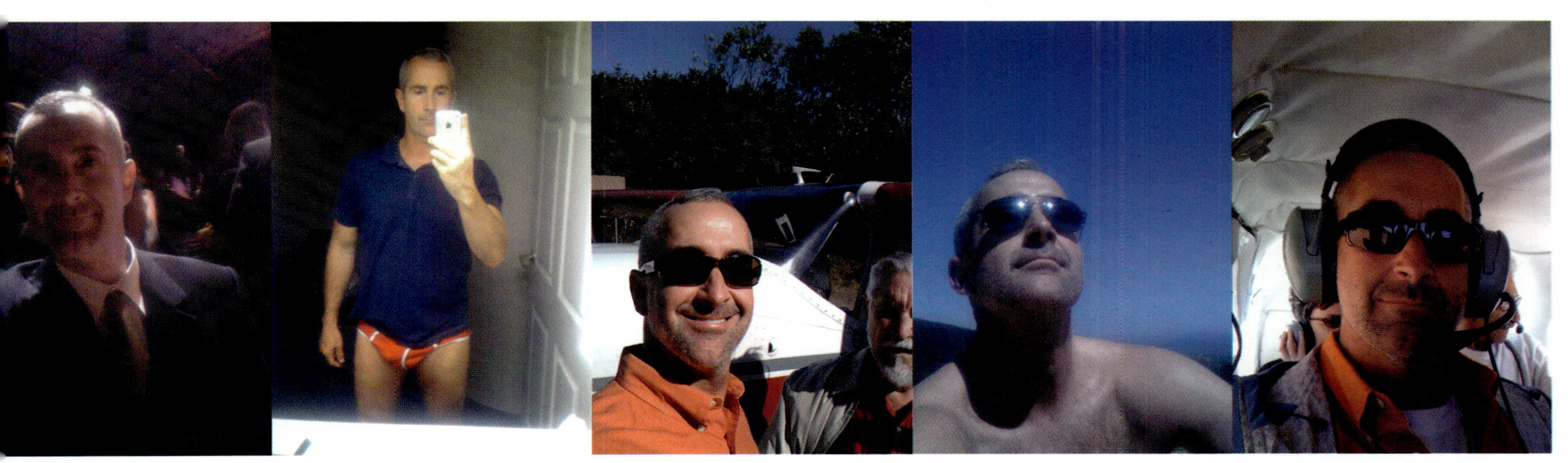

PETS & ANIMALS

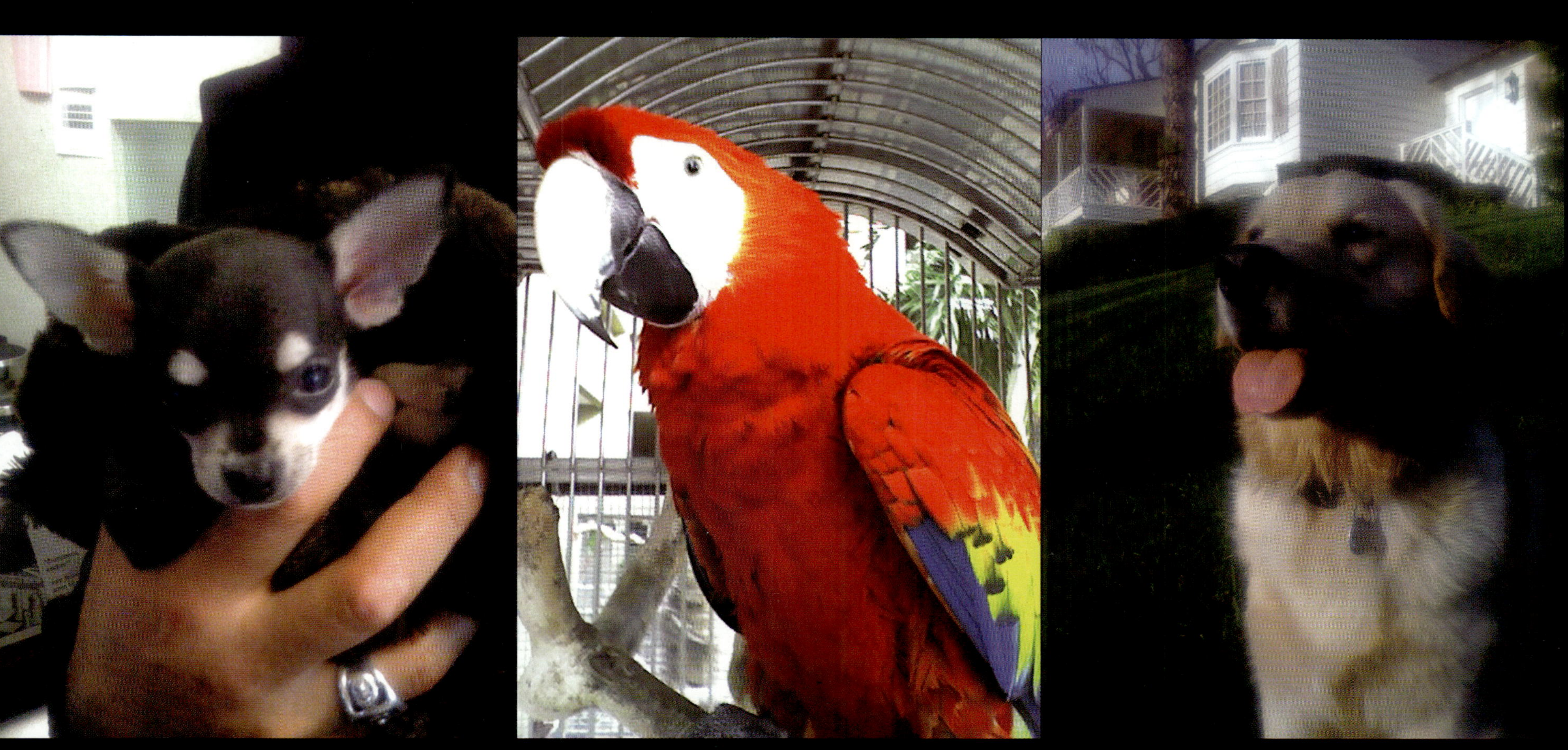

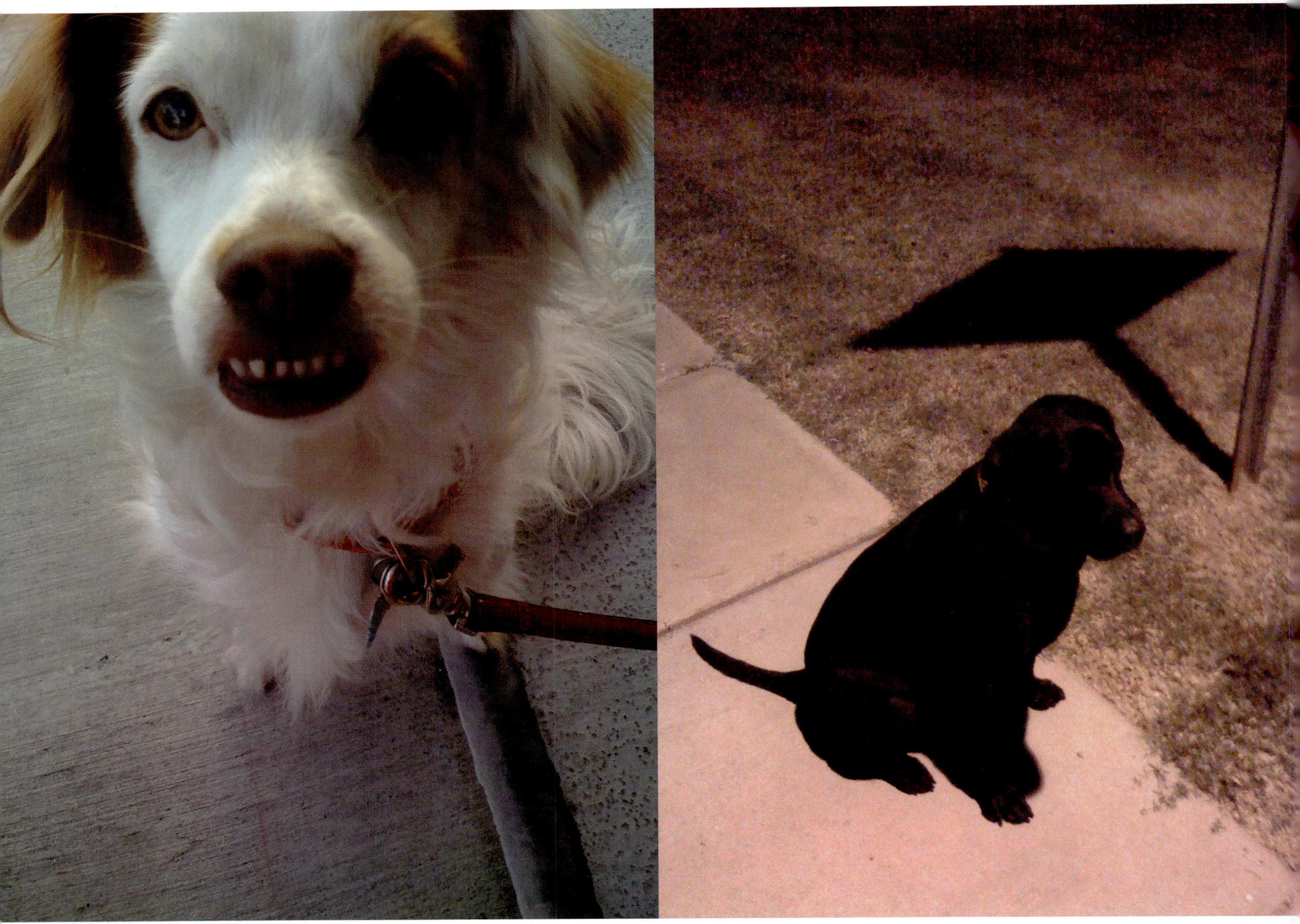

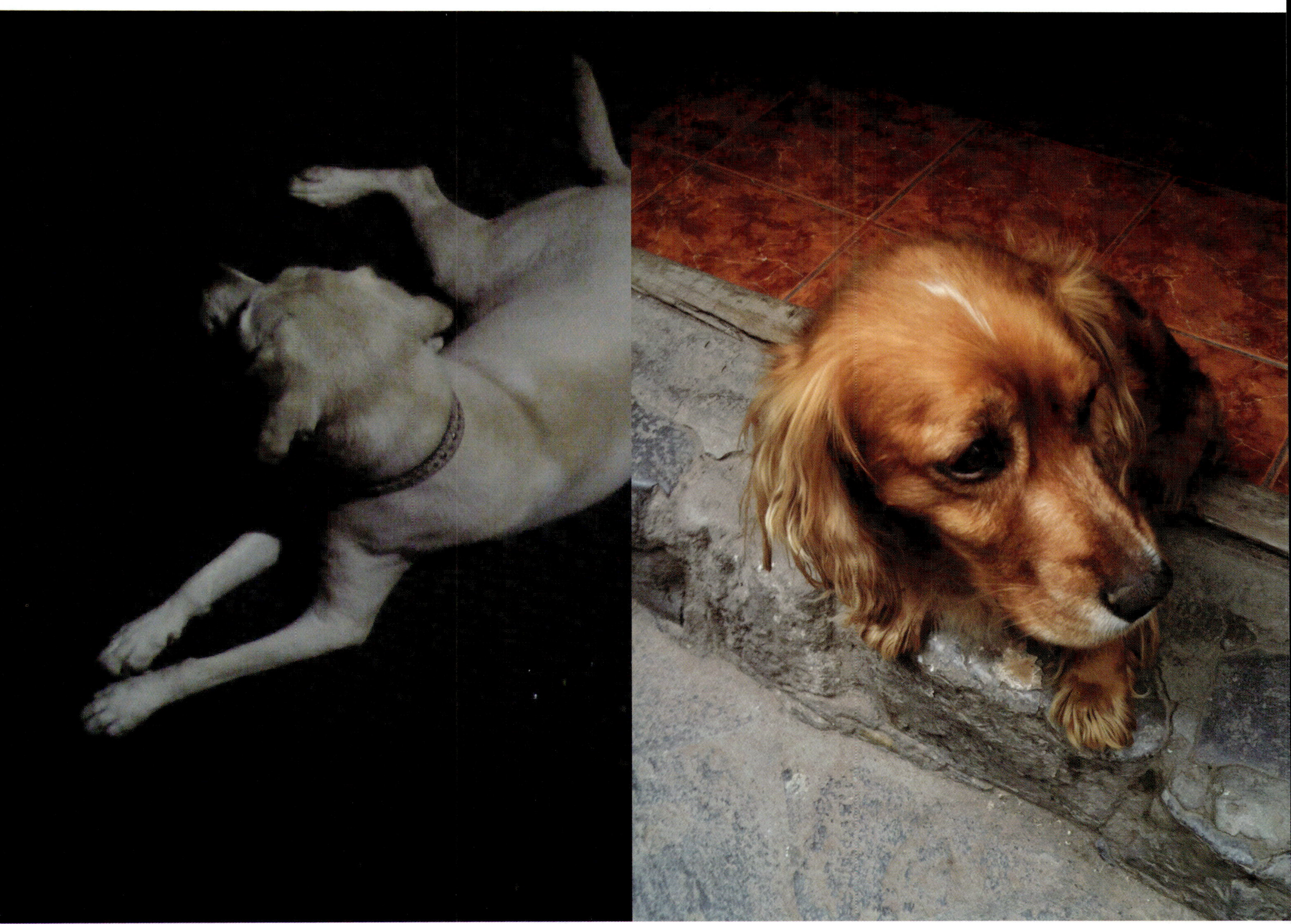

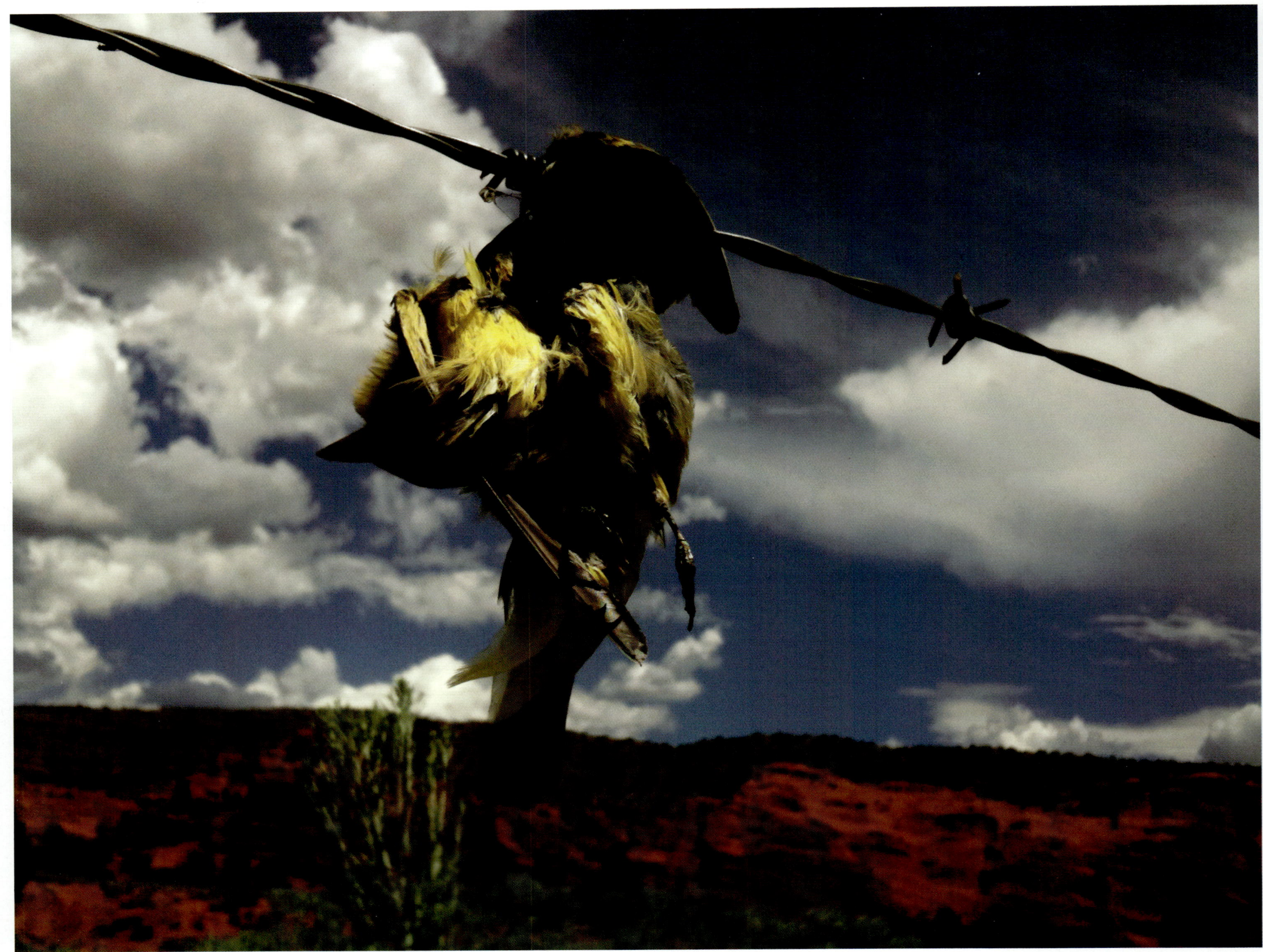

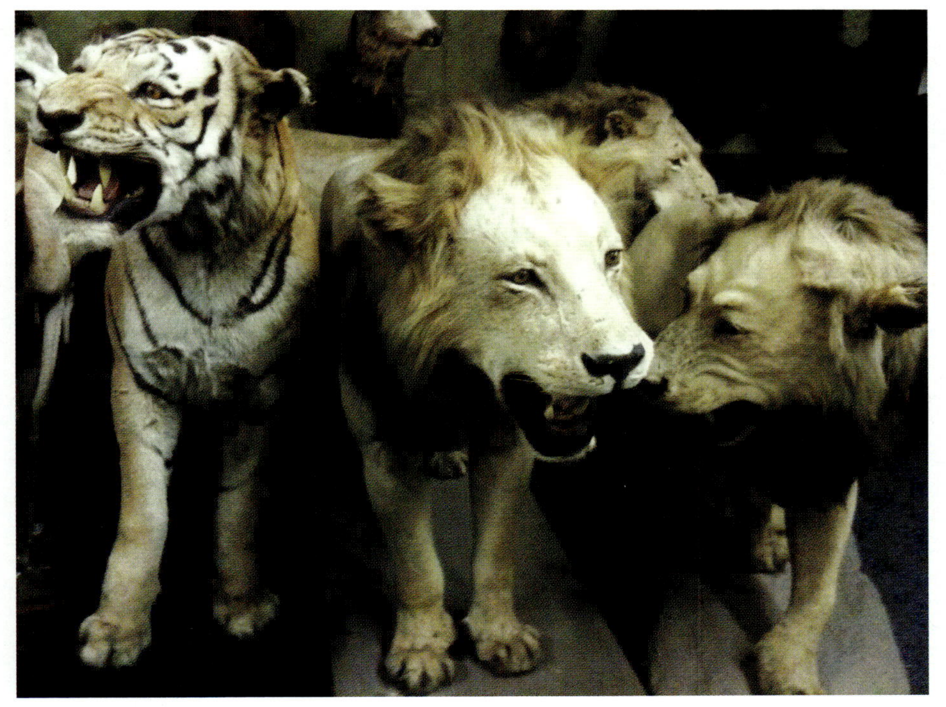

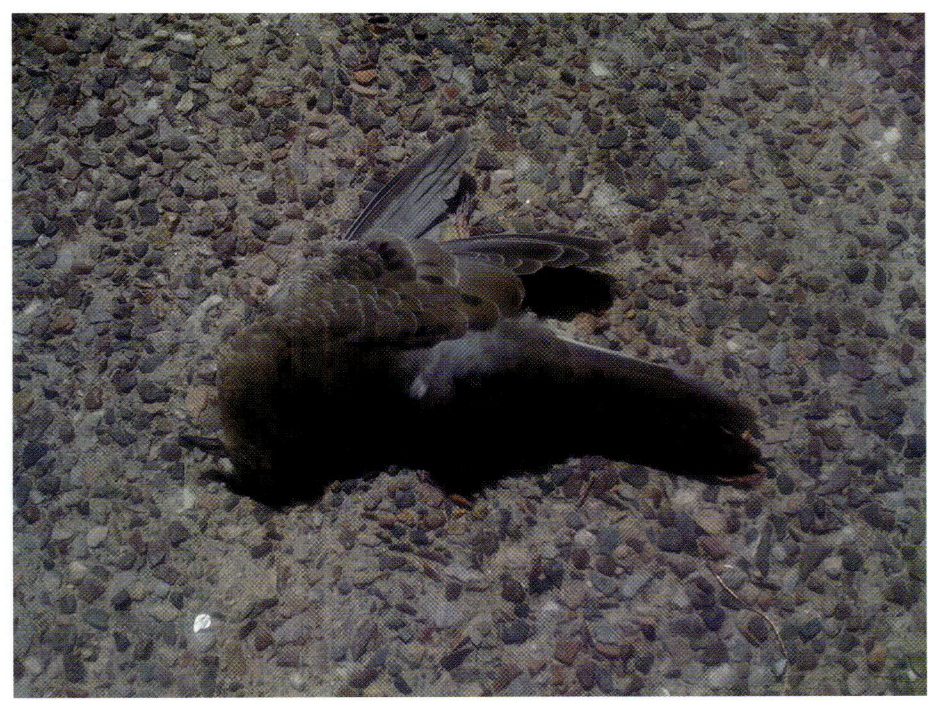

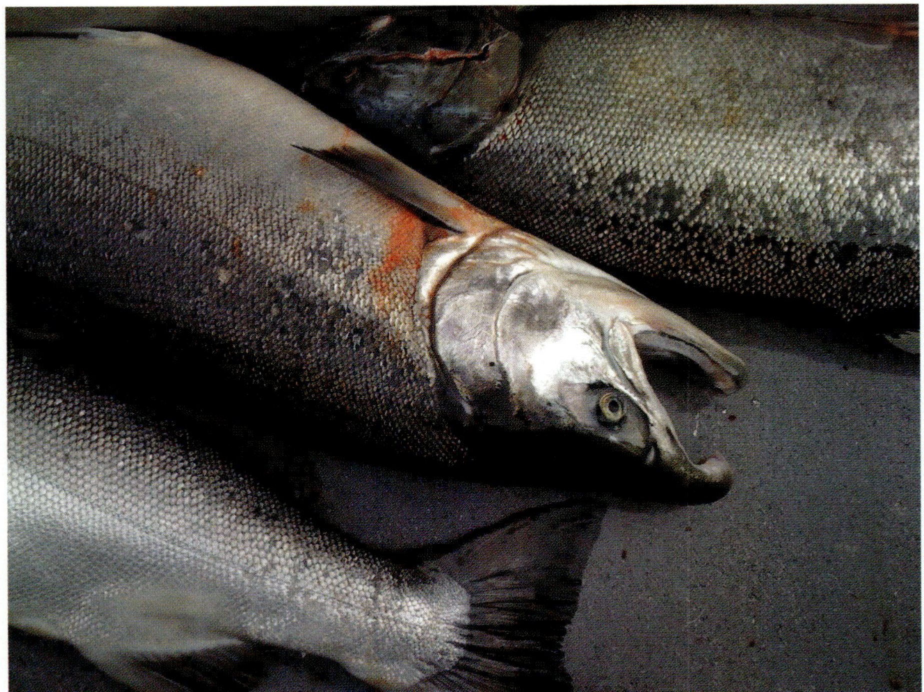

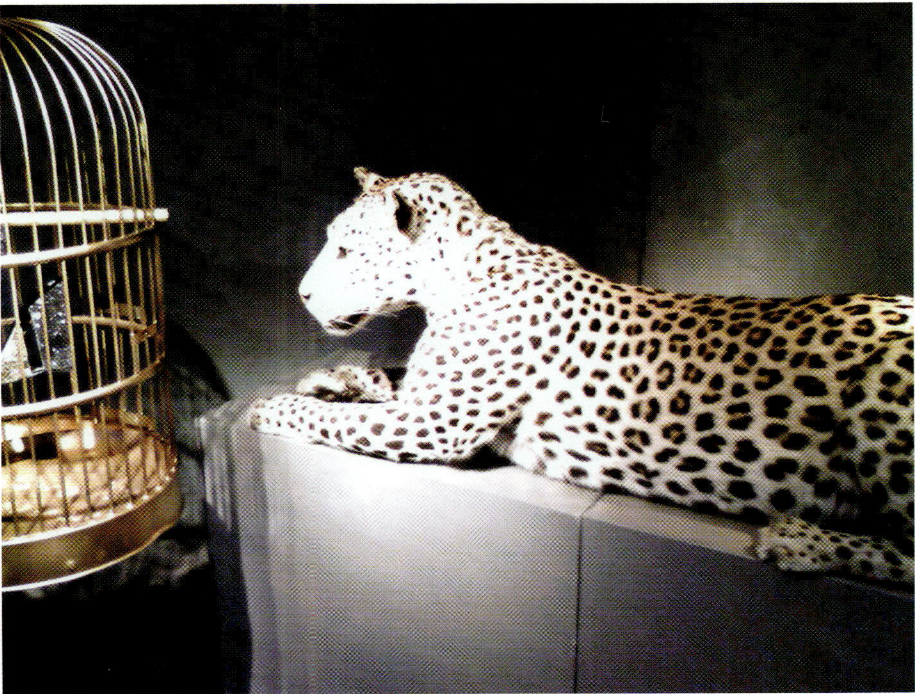

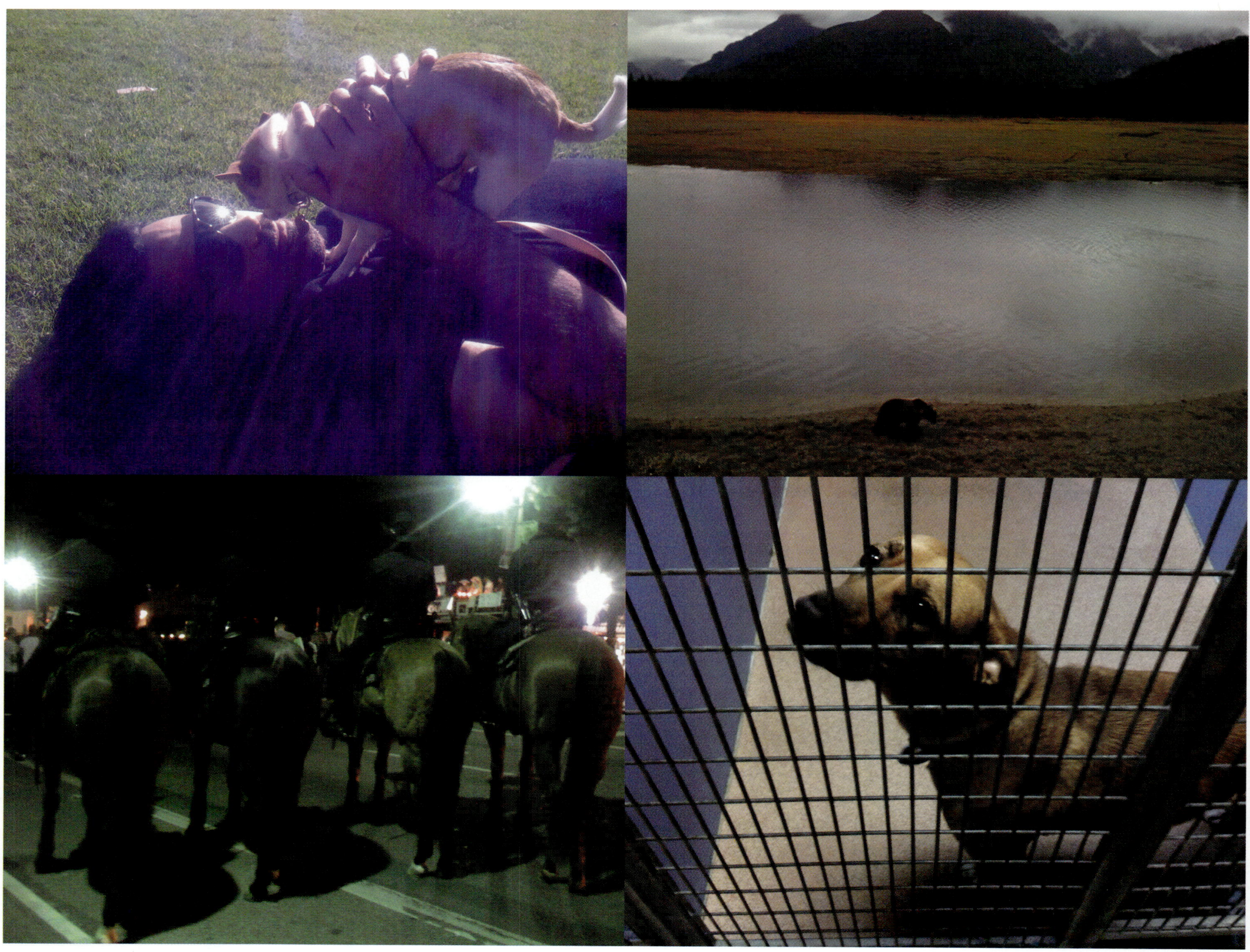

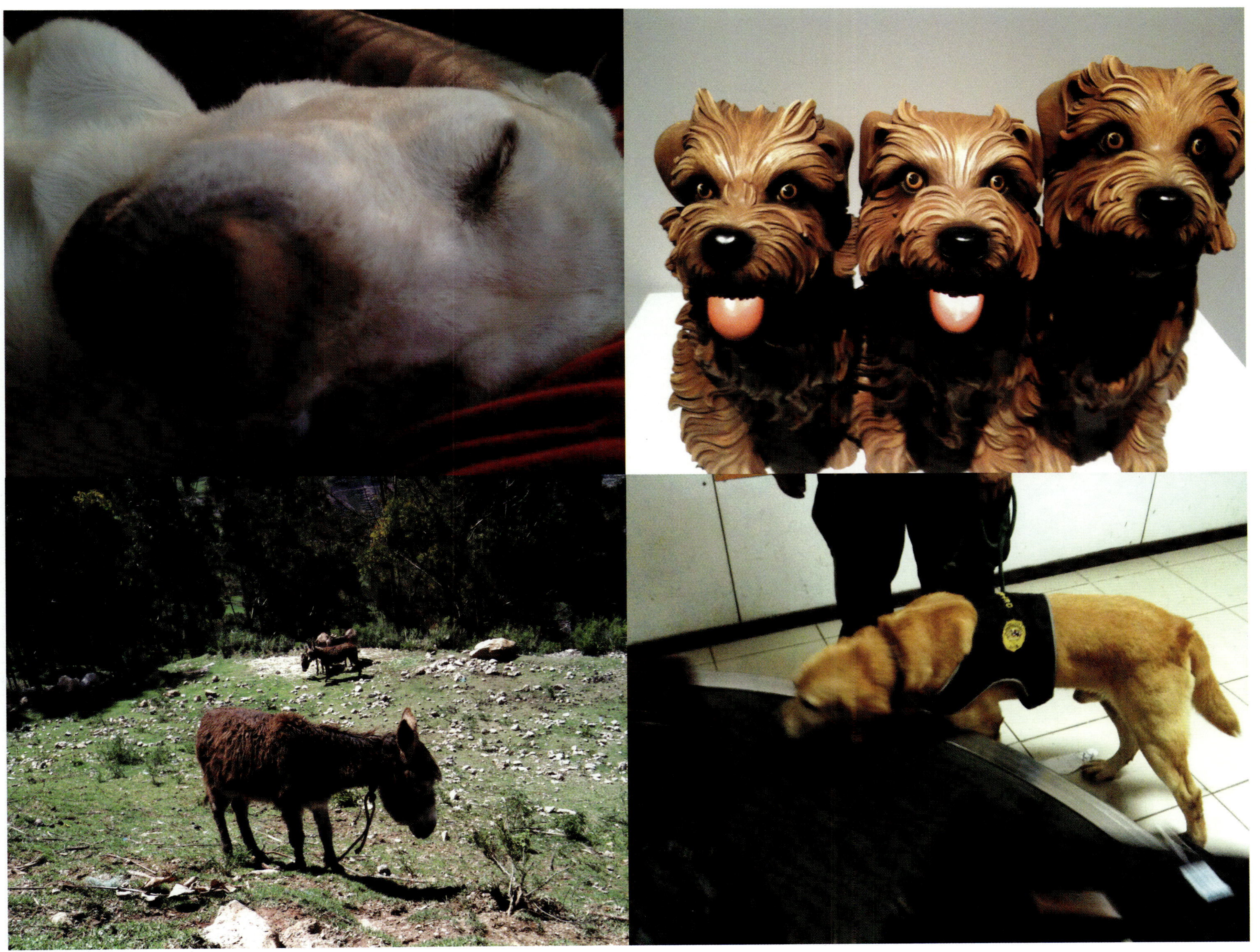

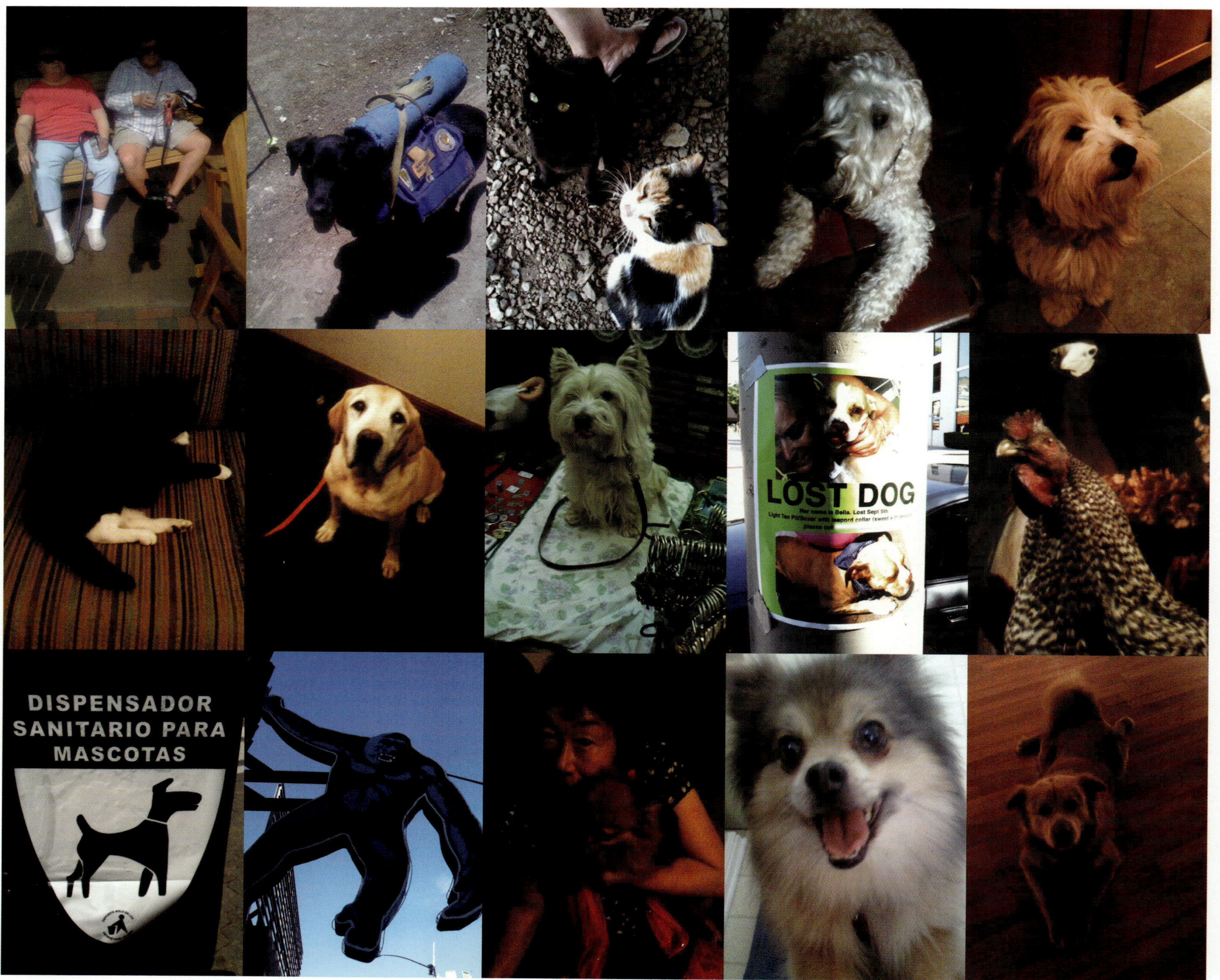
DISPENSADOR
SANITARIO PARA
MASCOTAS

LOST DOG

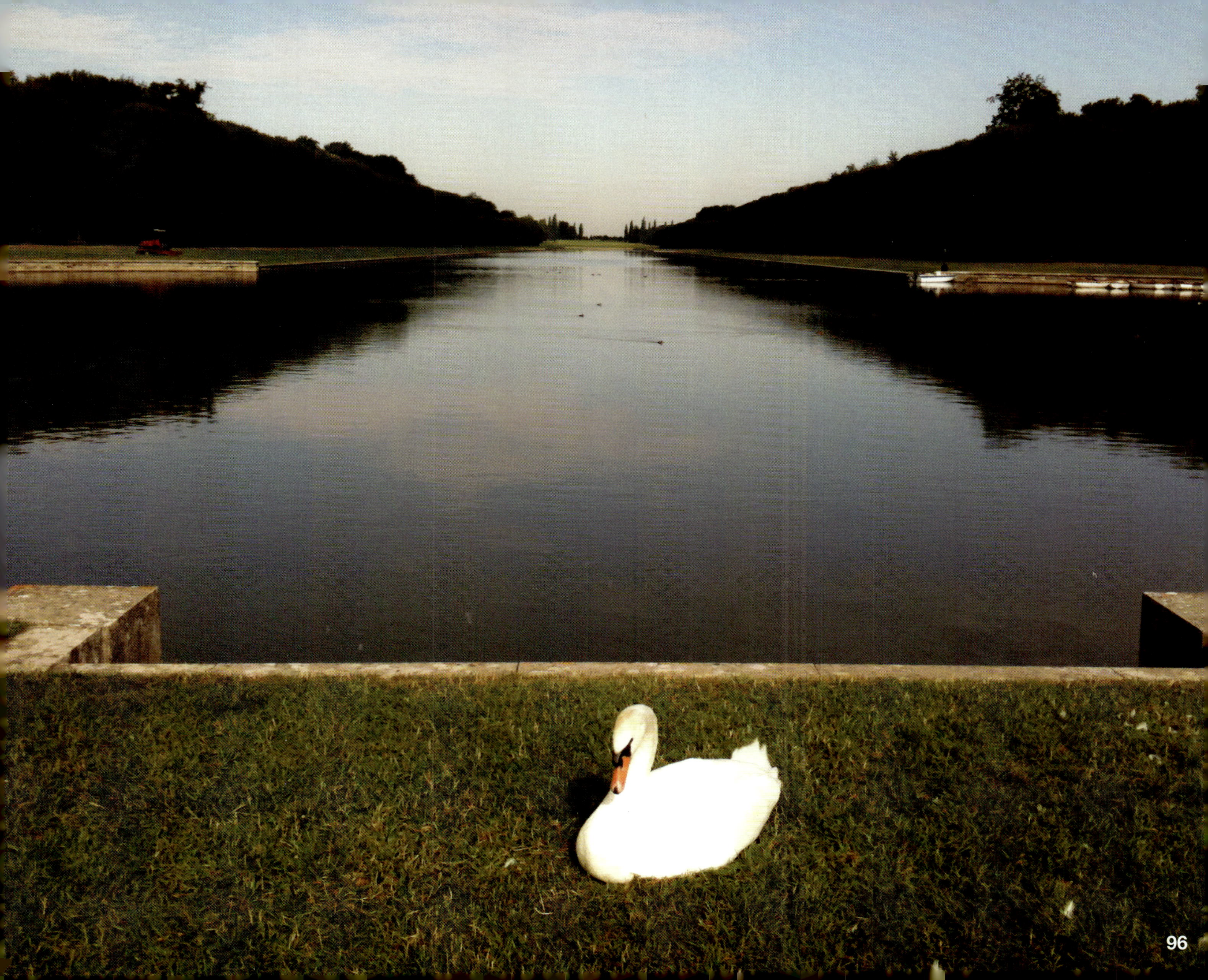

TRANSPORTATION

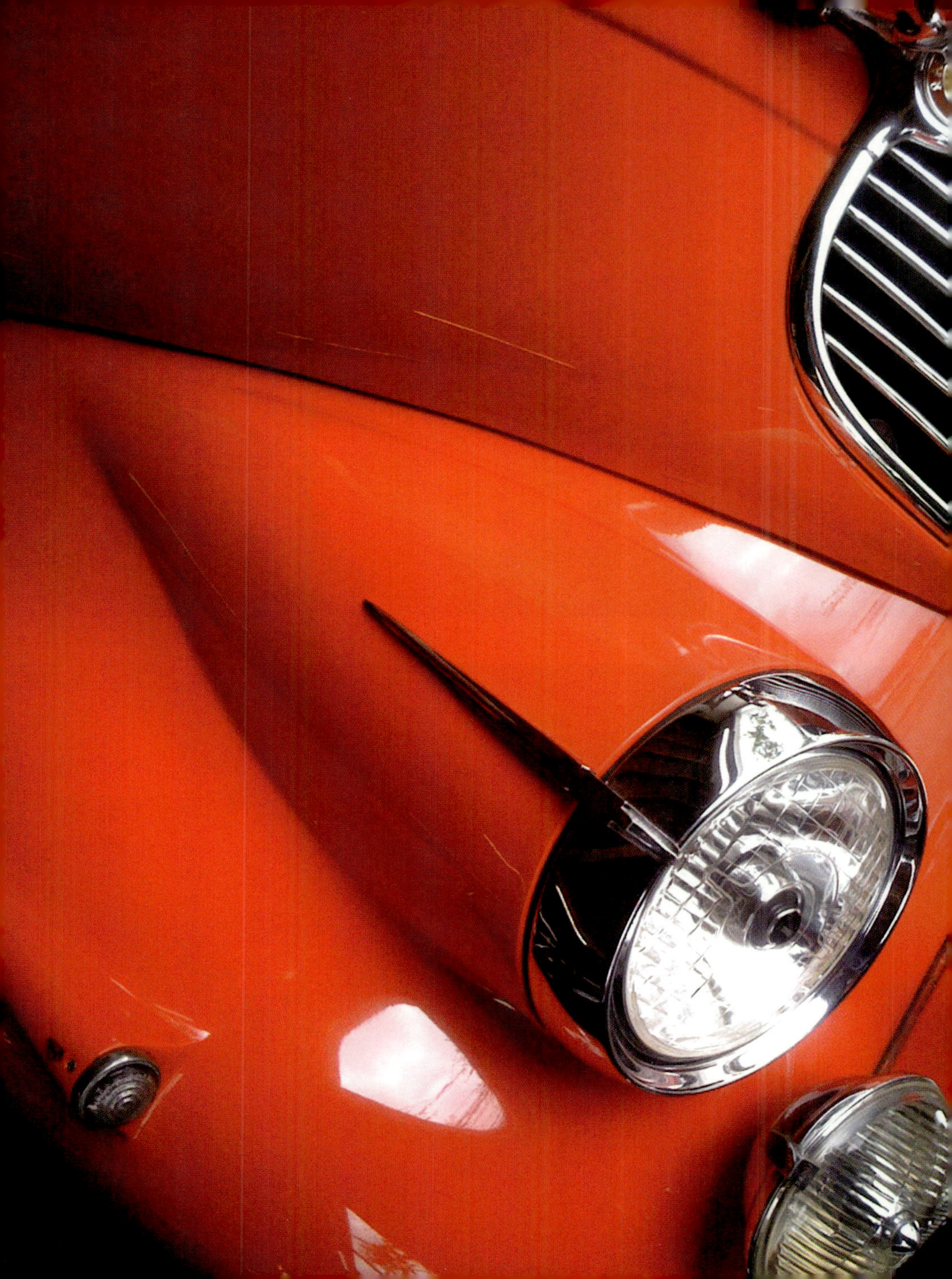

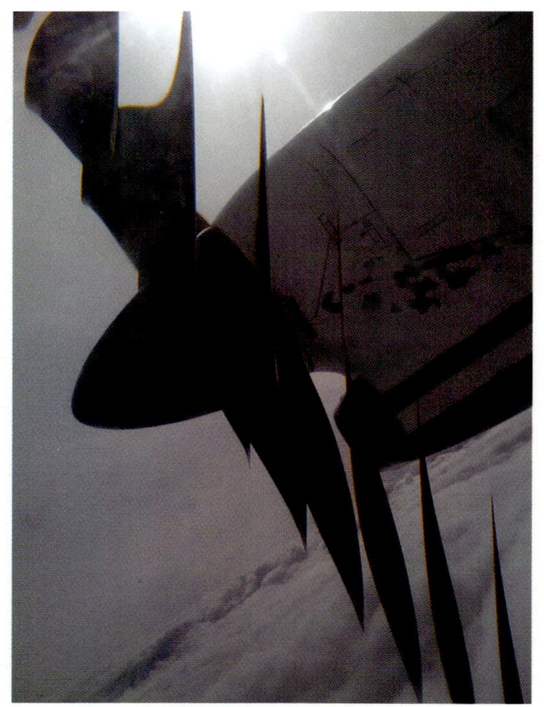
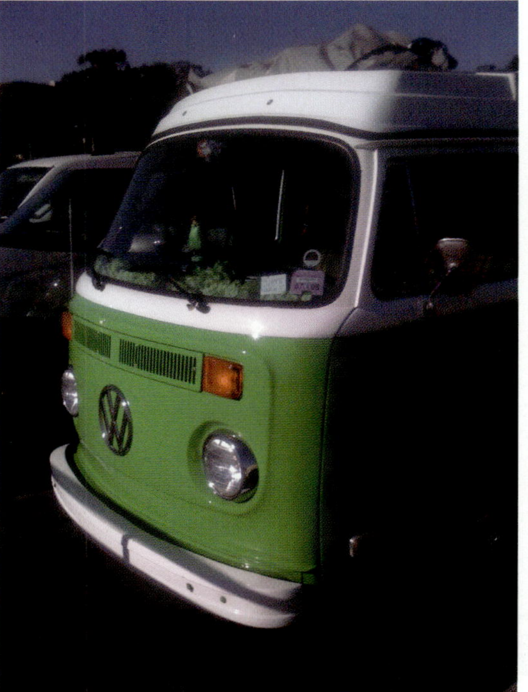
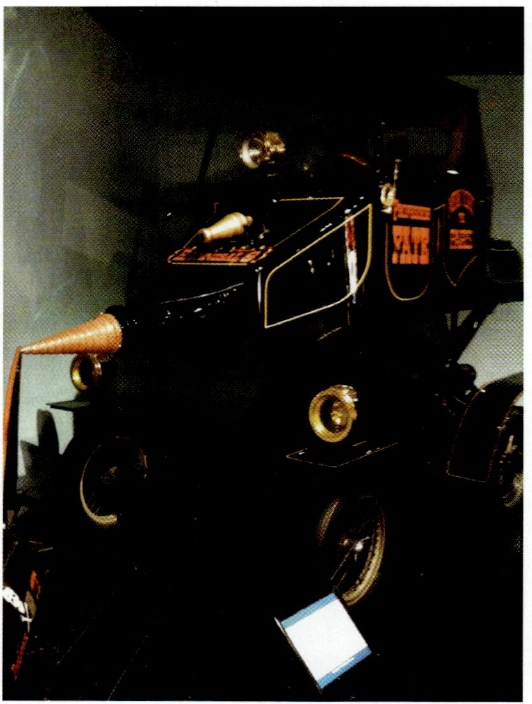
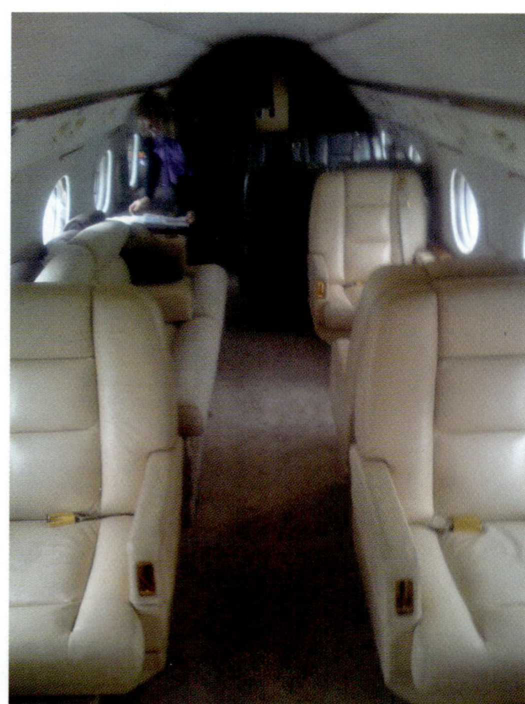
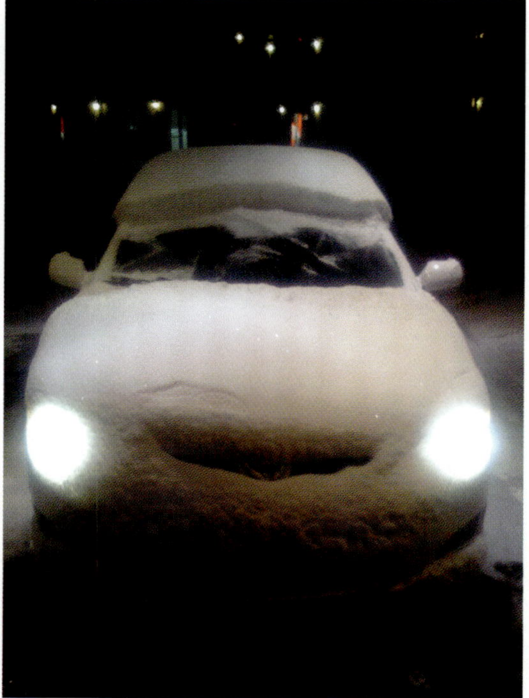
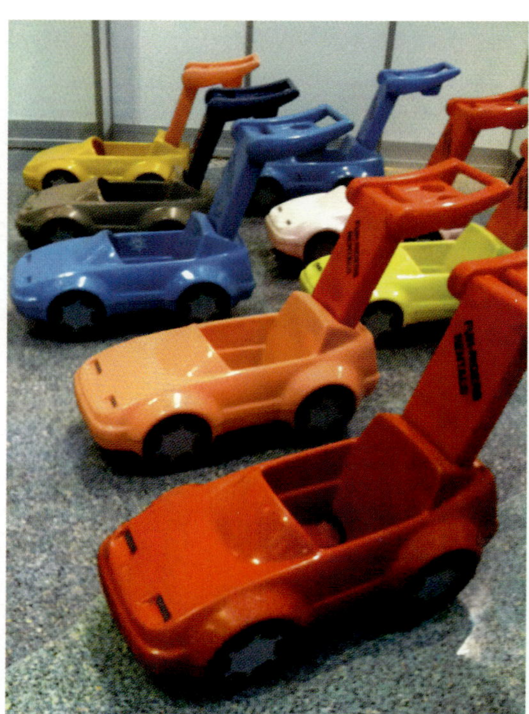

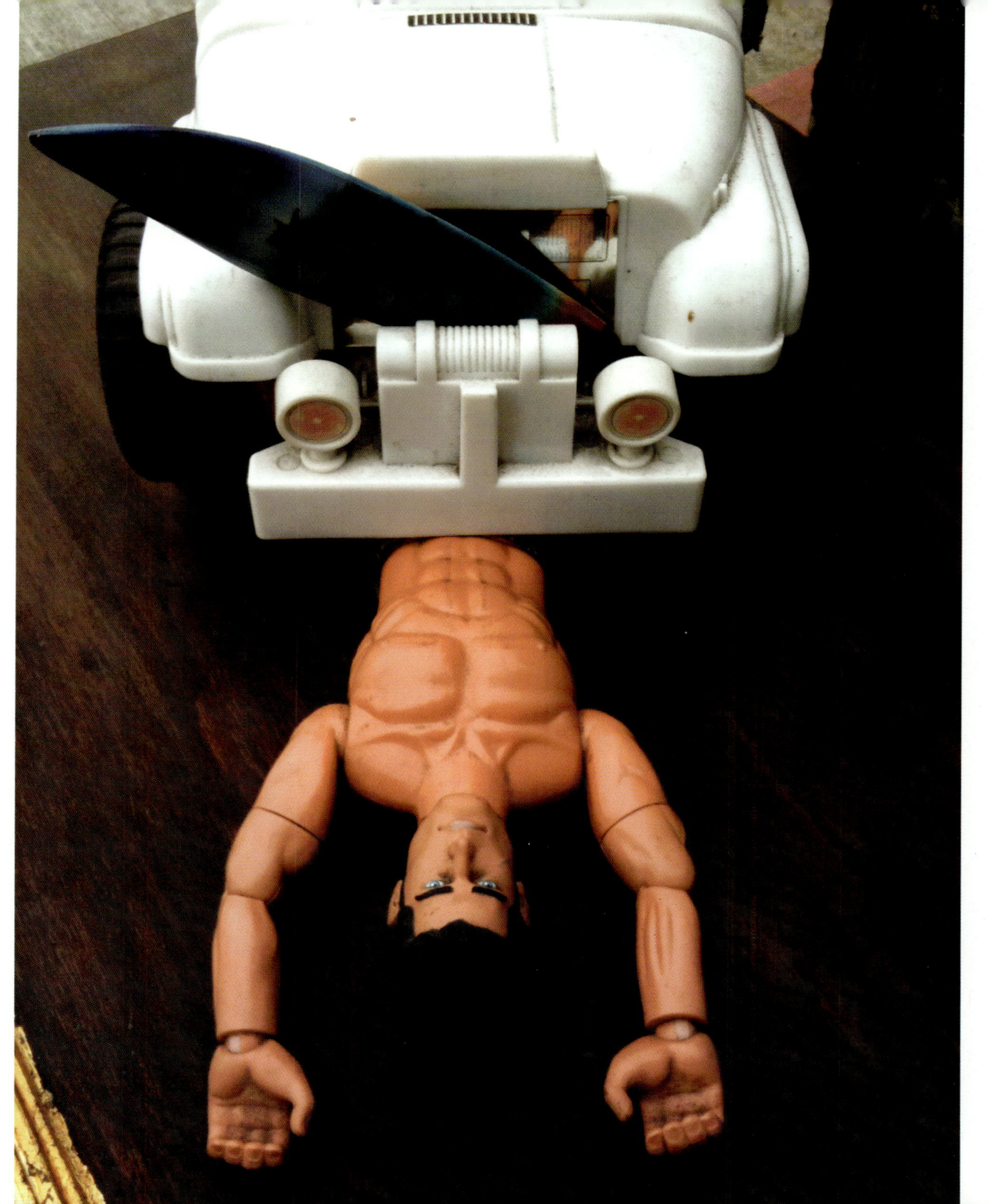

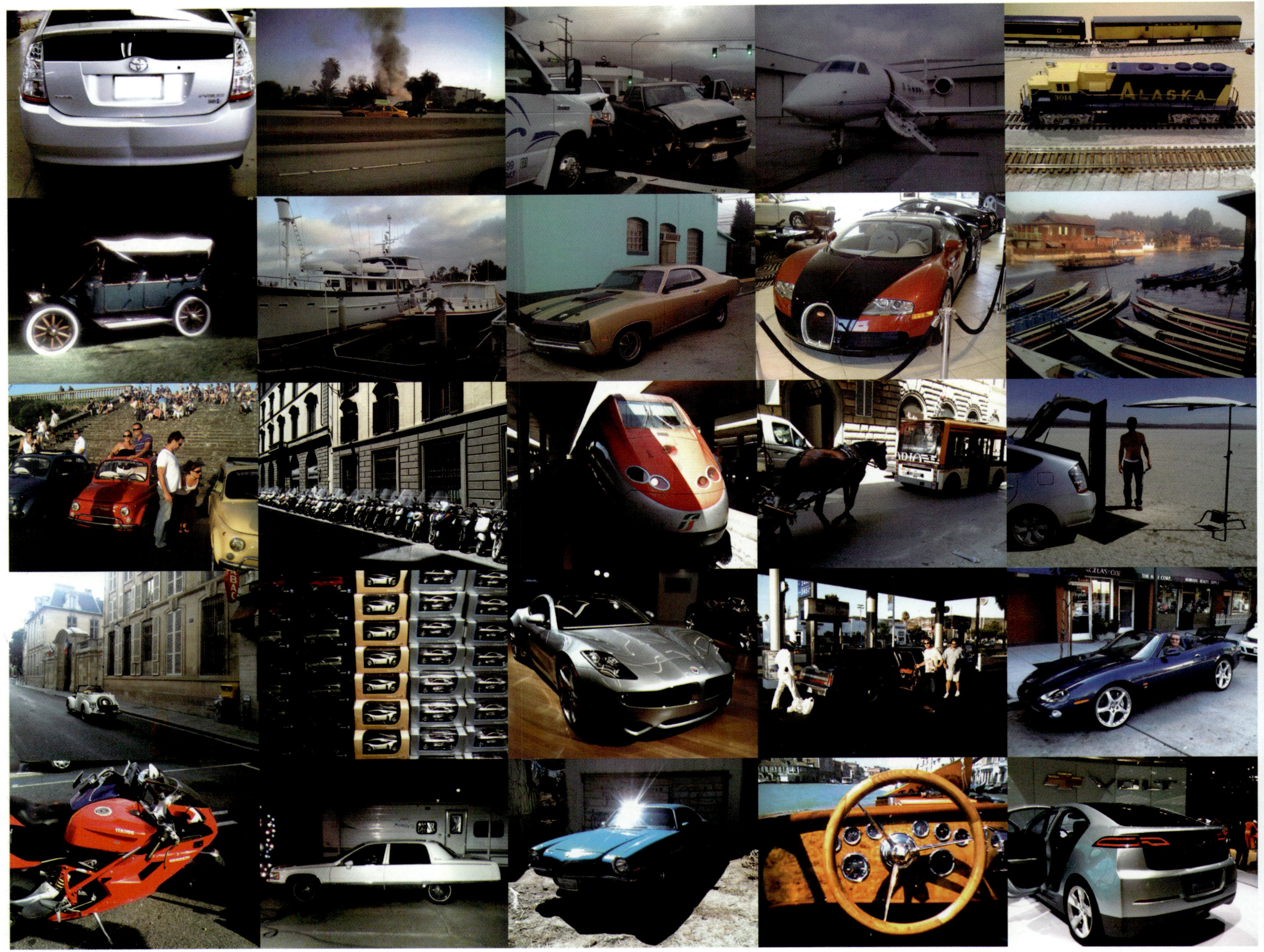

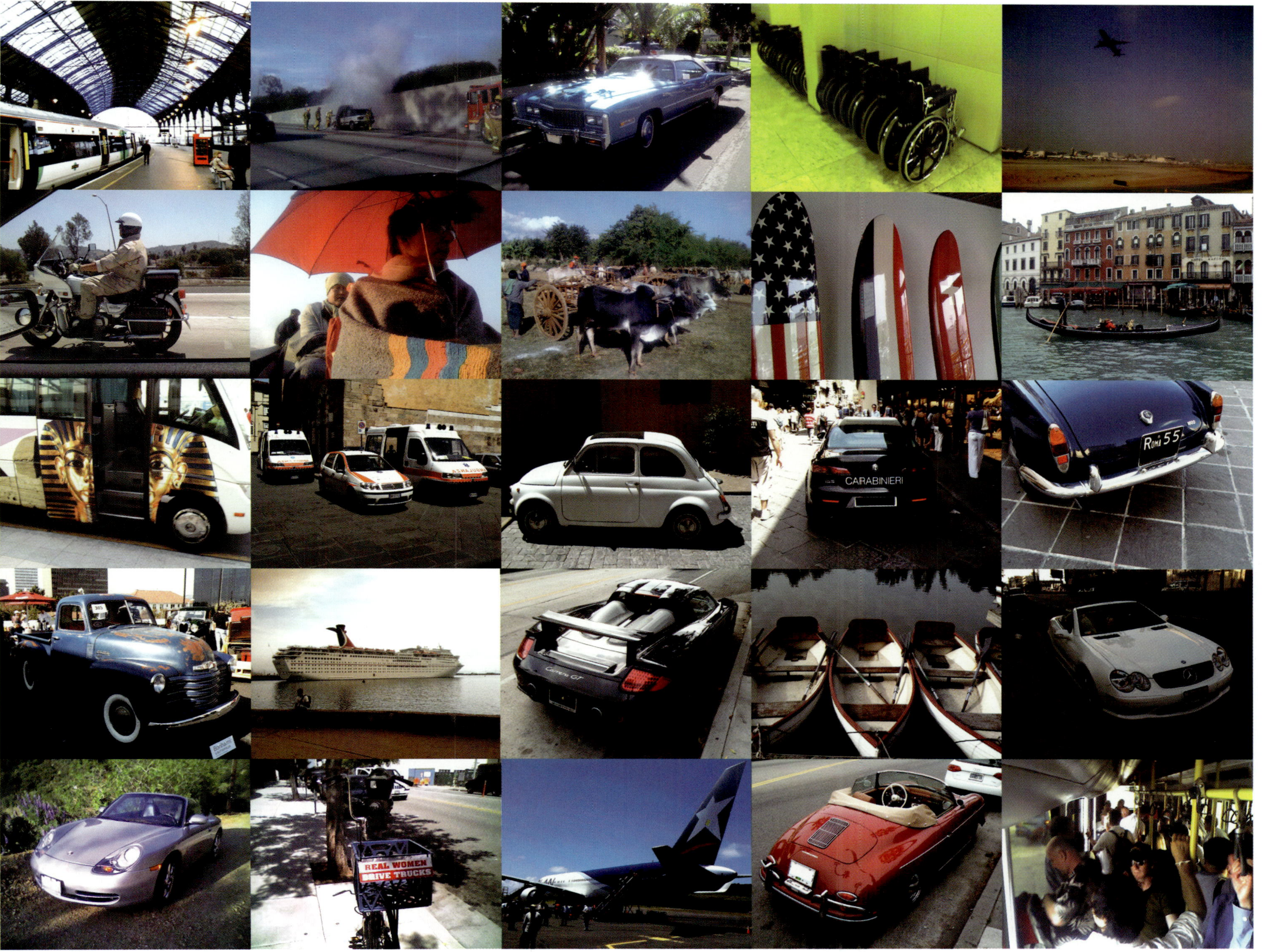

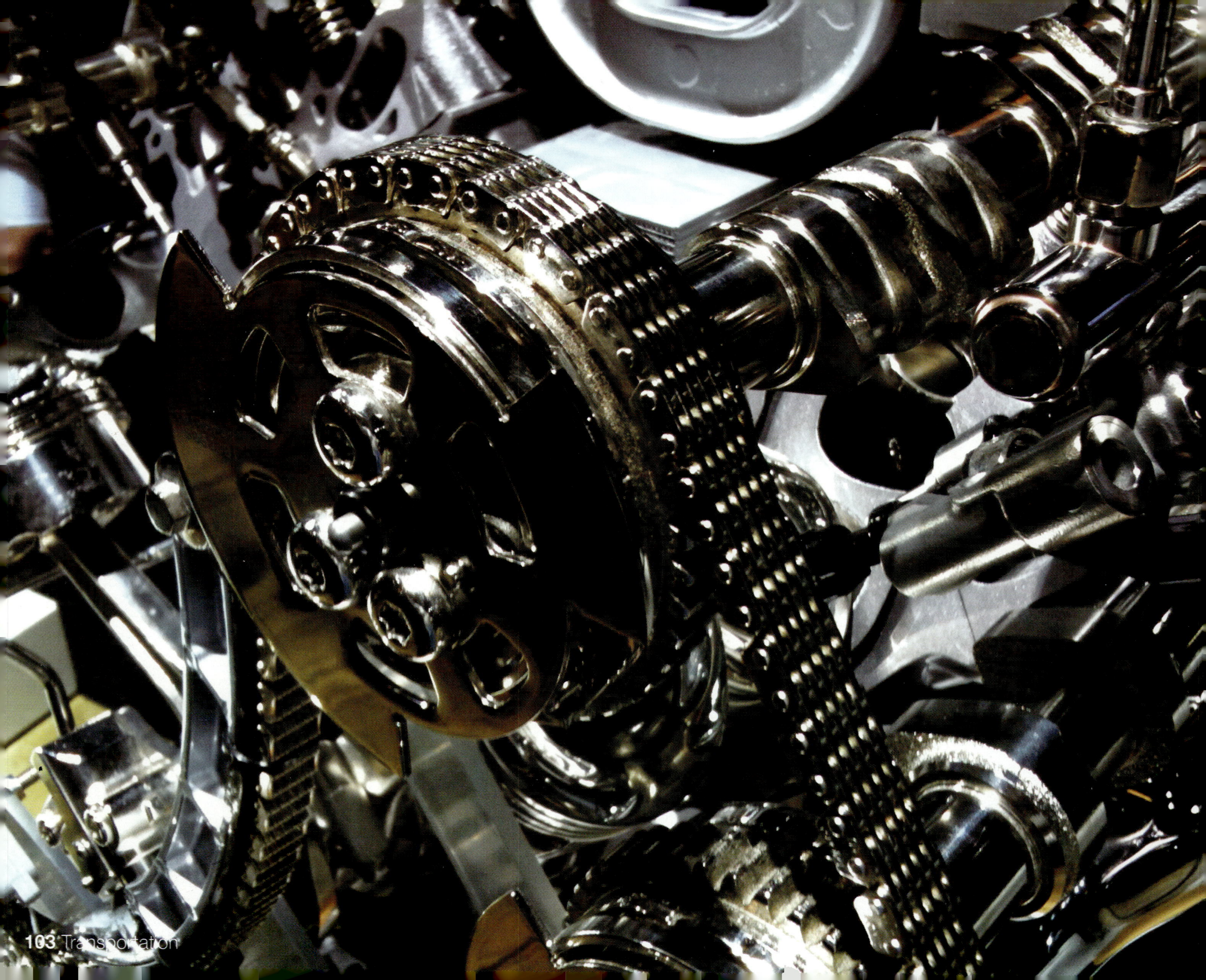

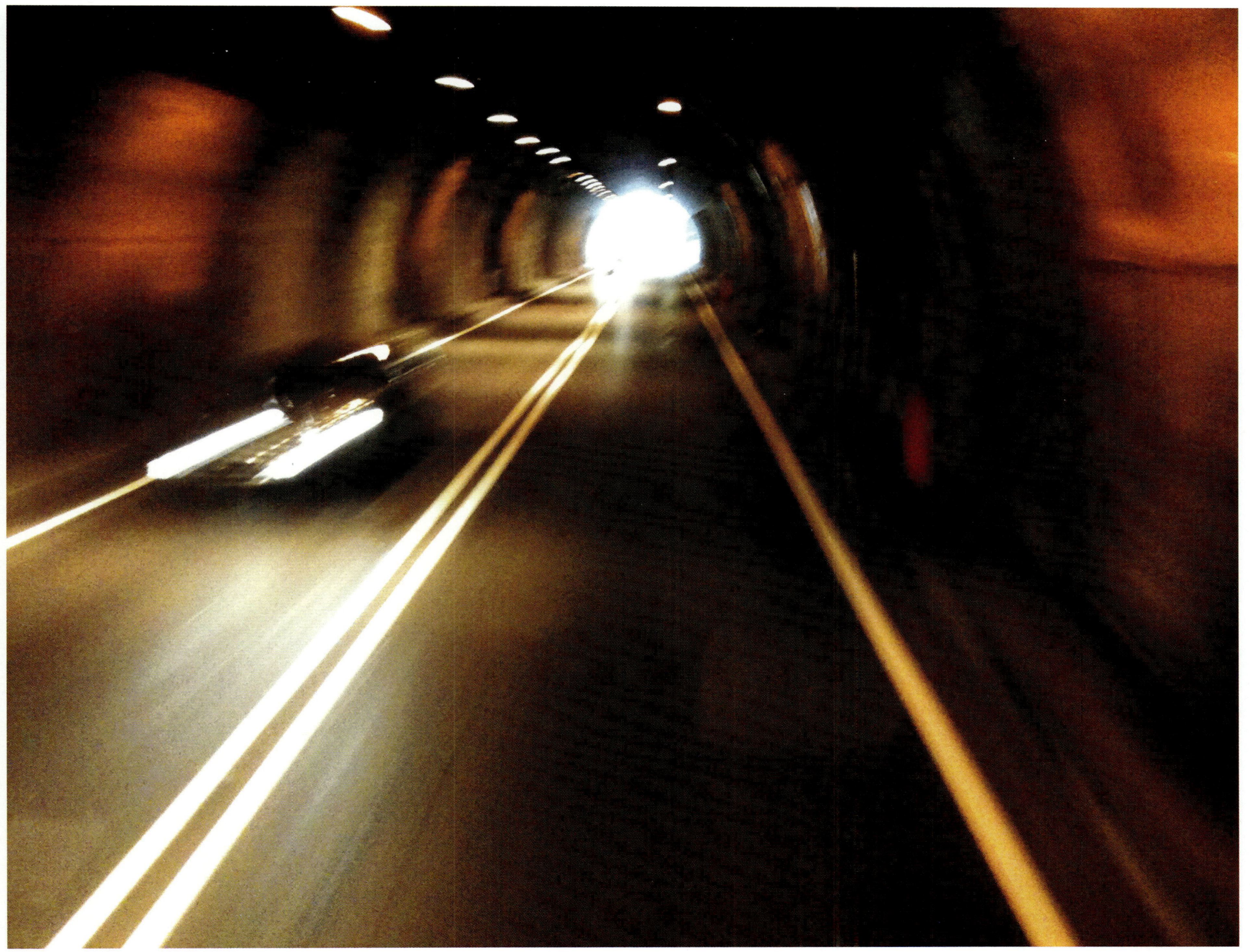

TRAVEL

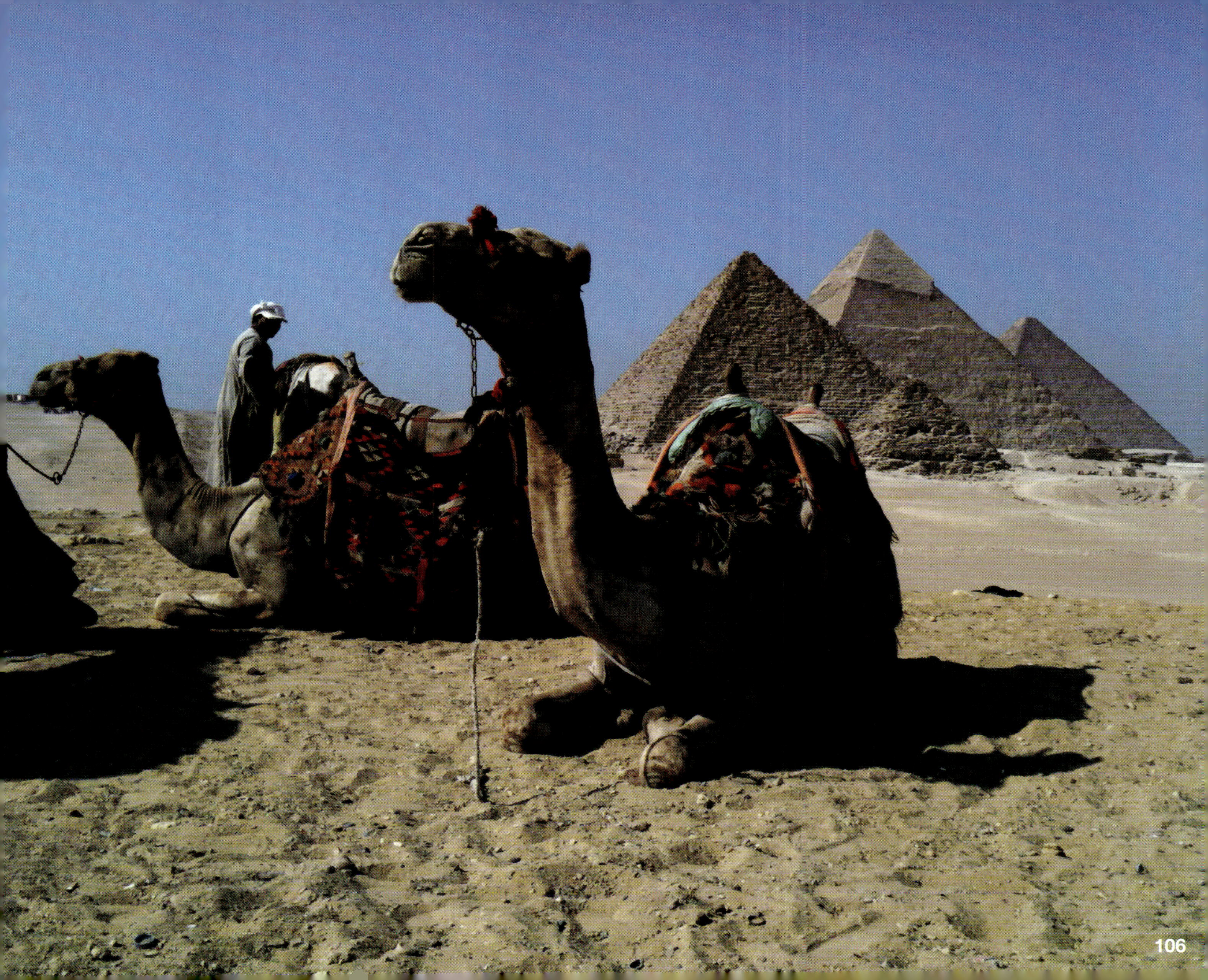

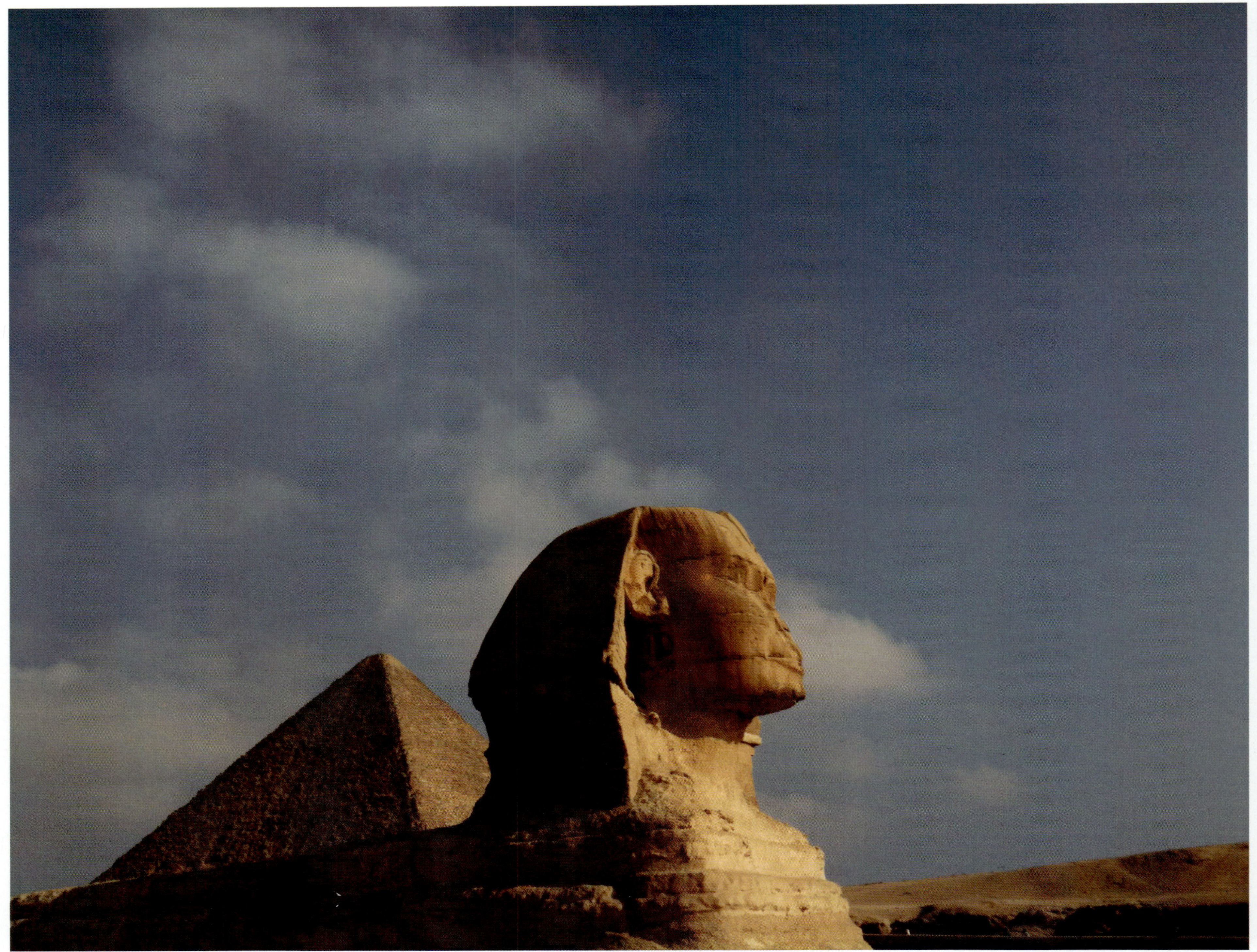

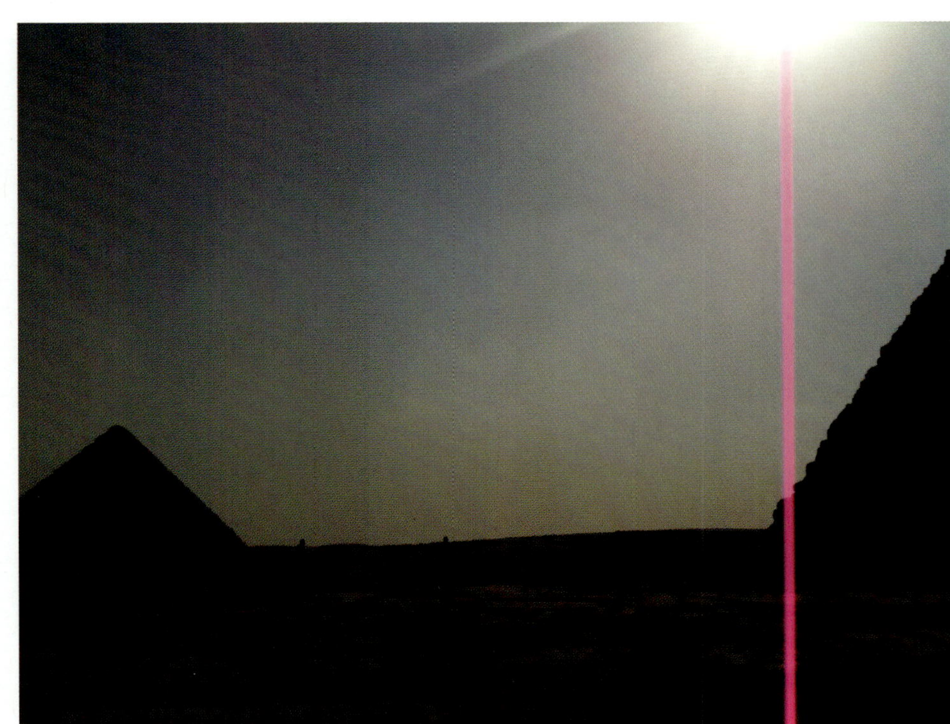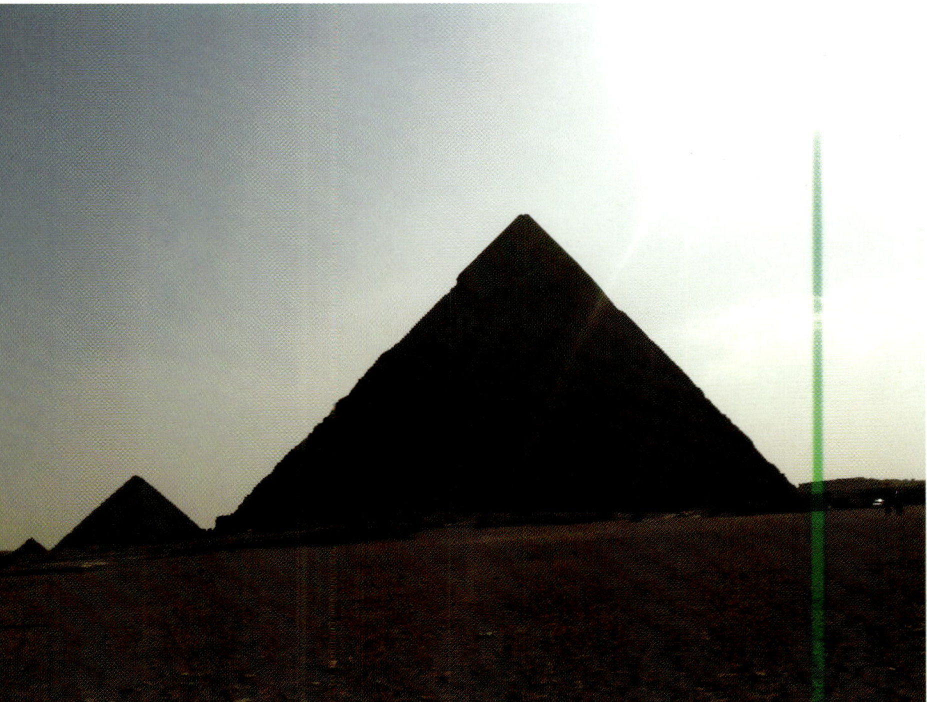

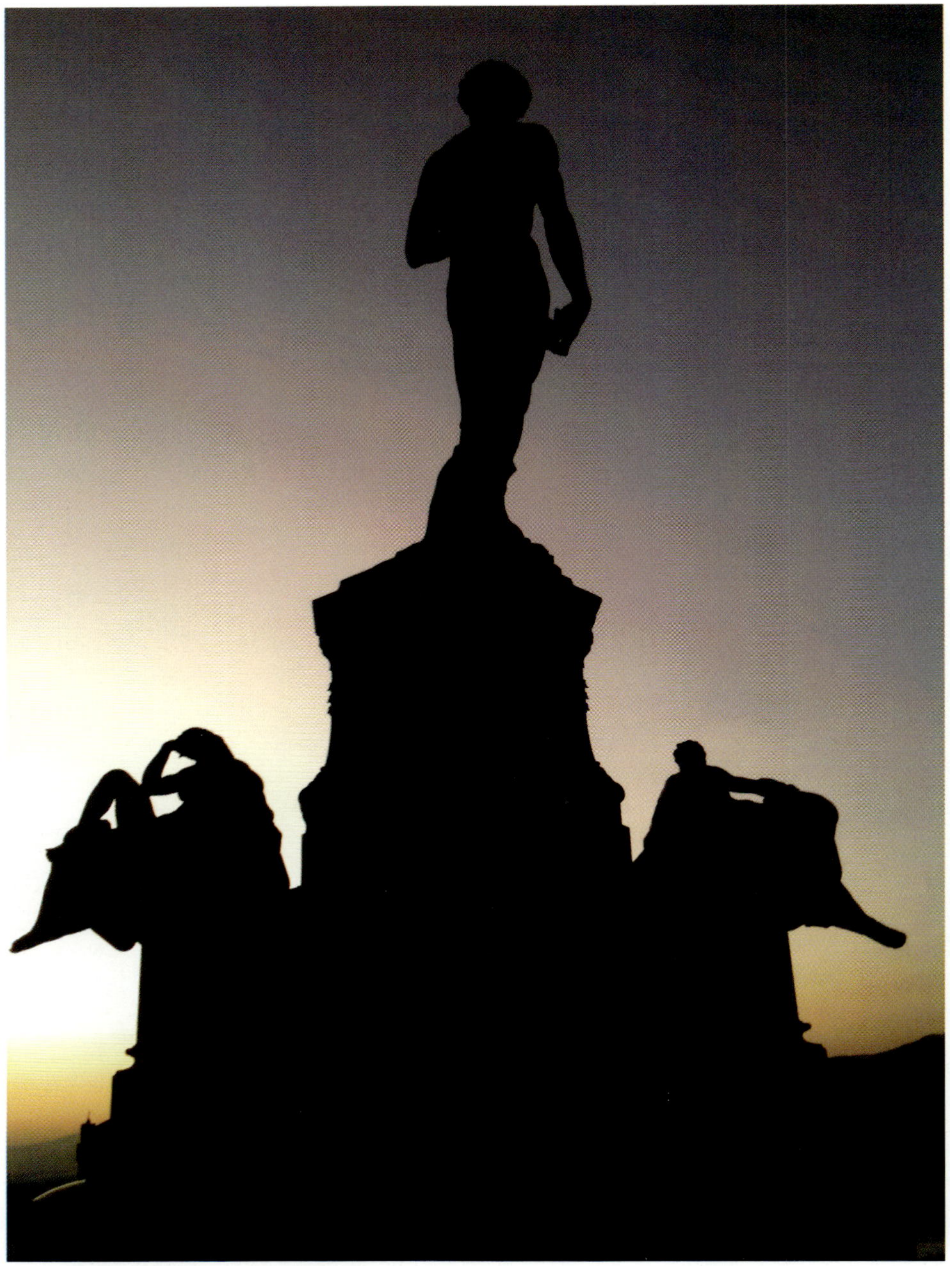

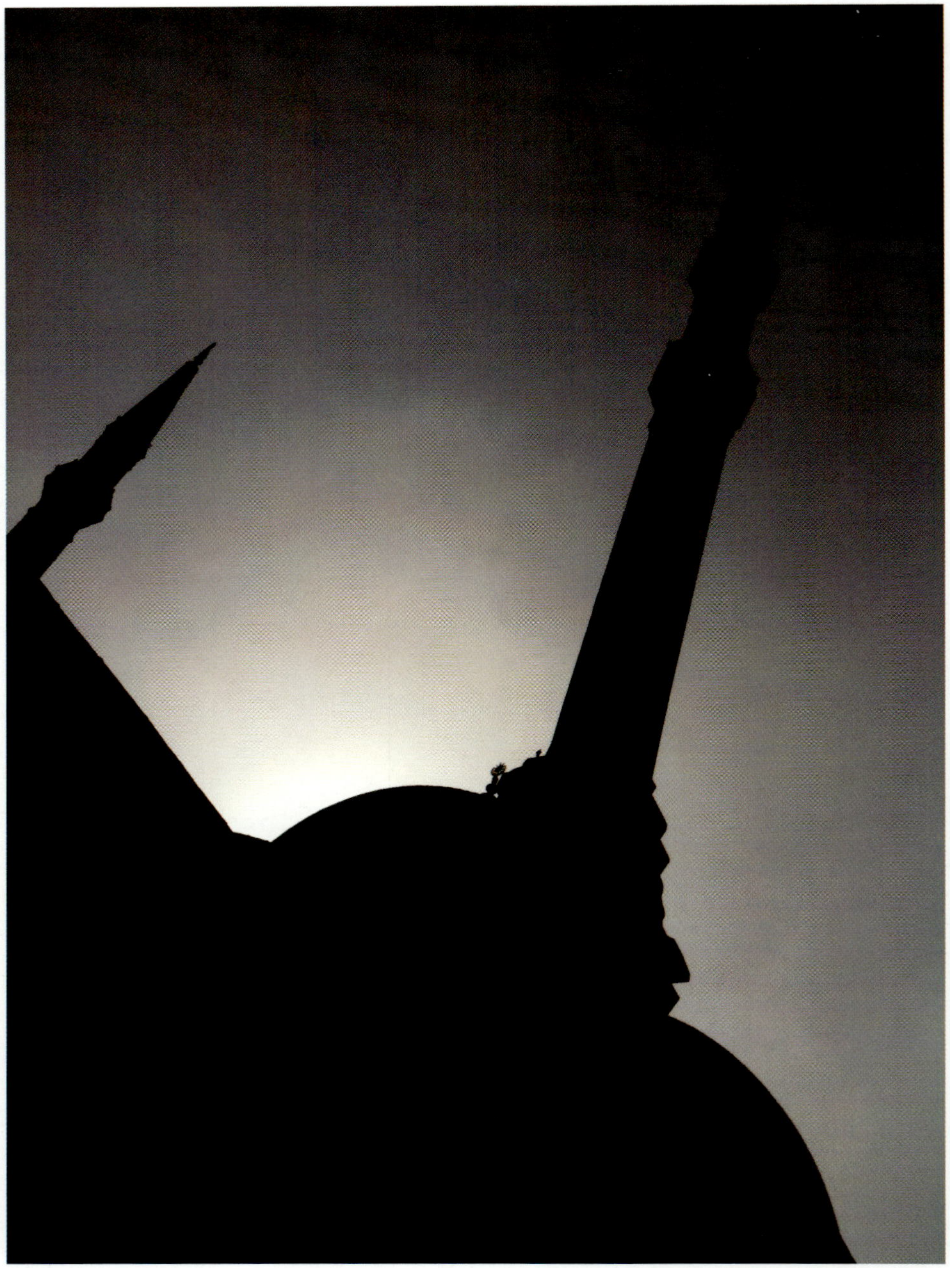

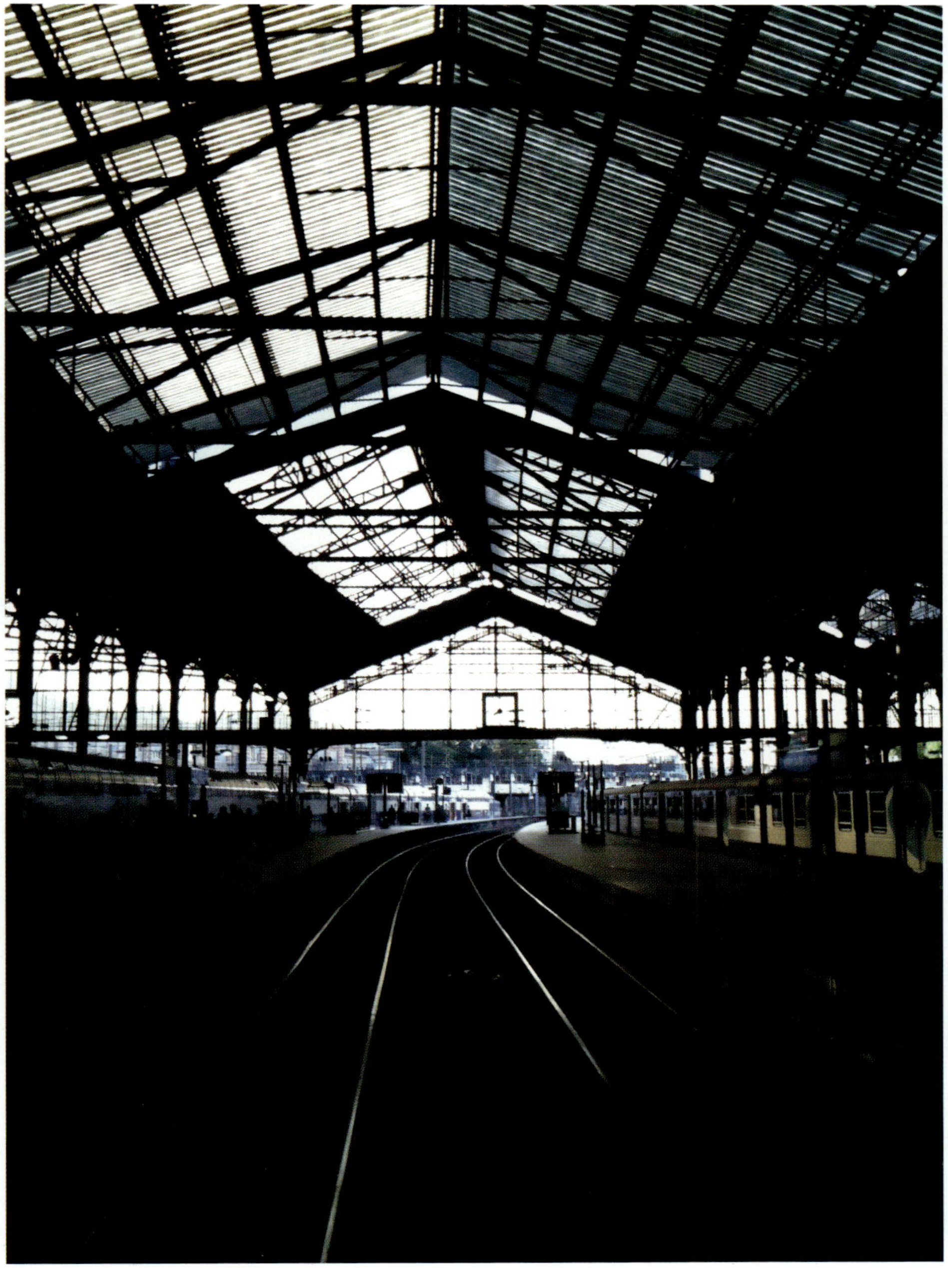

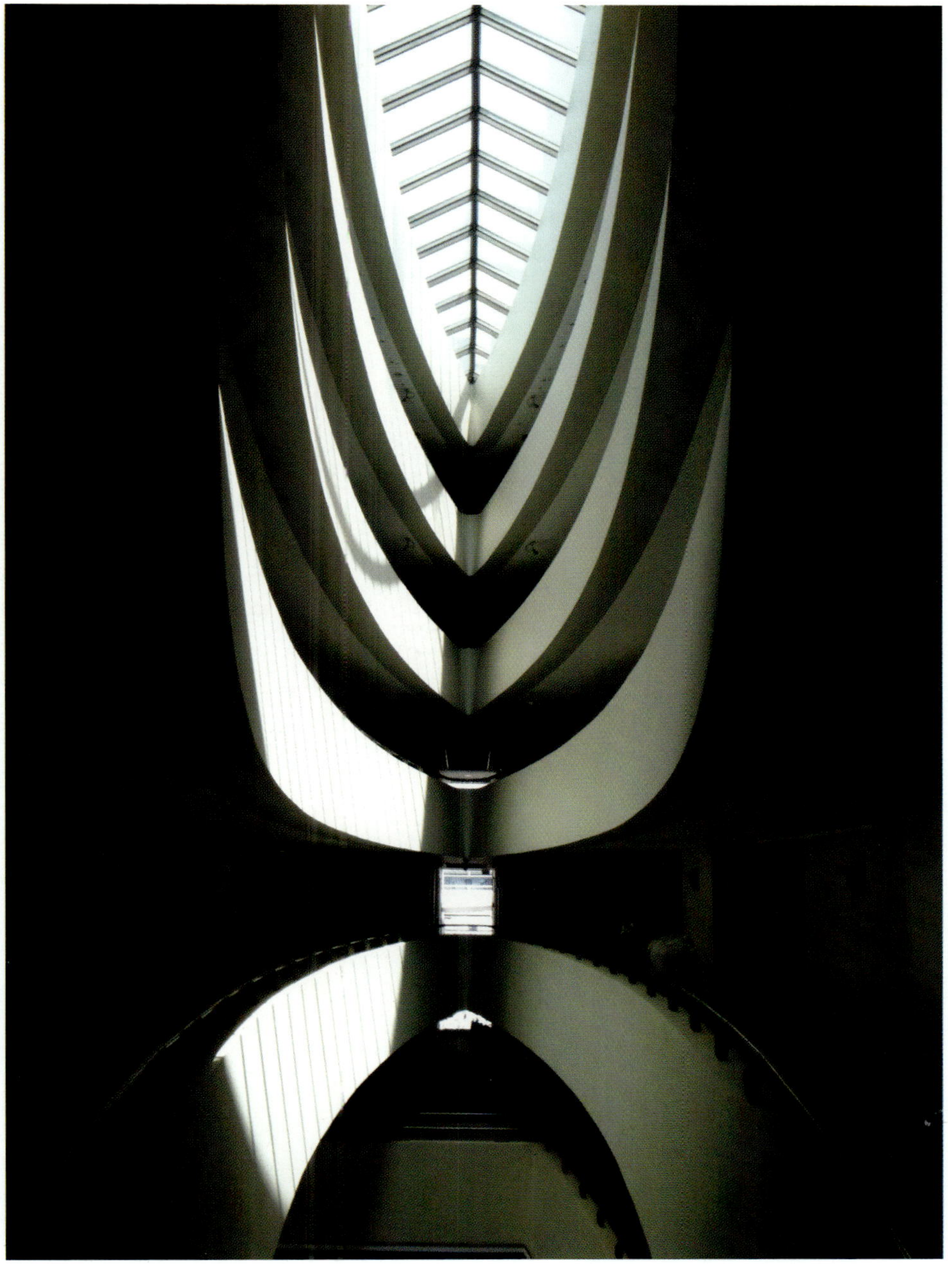

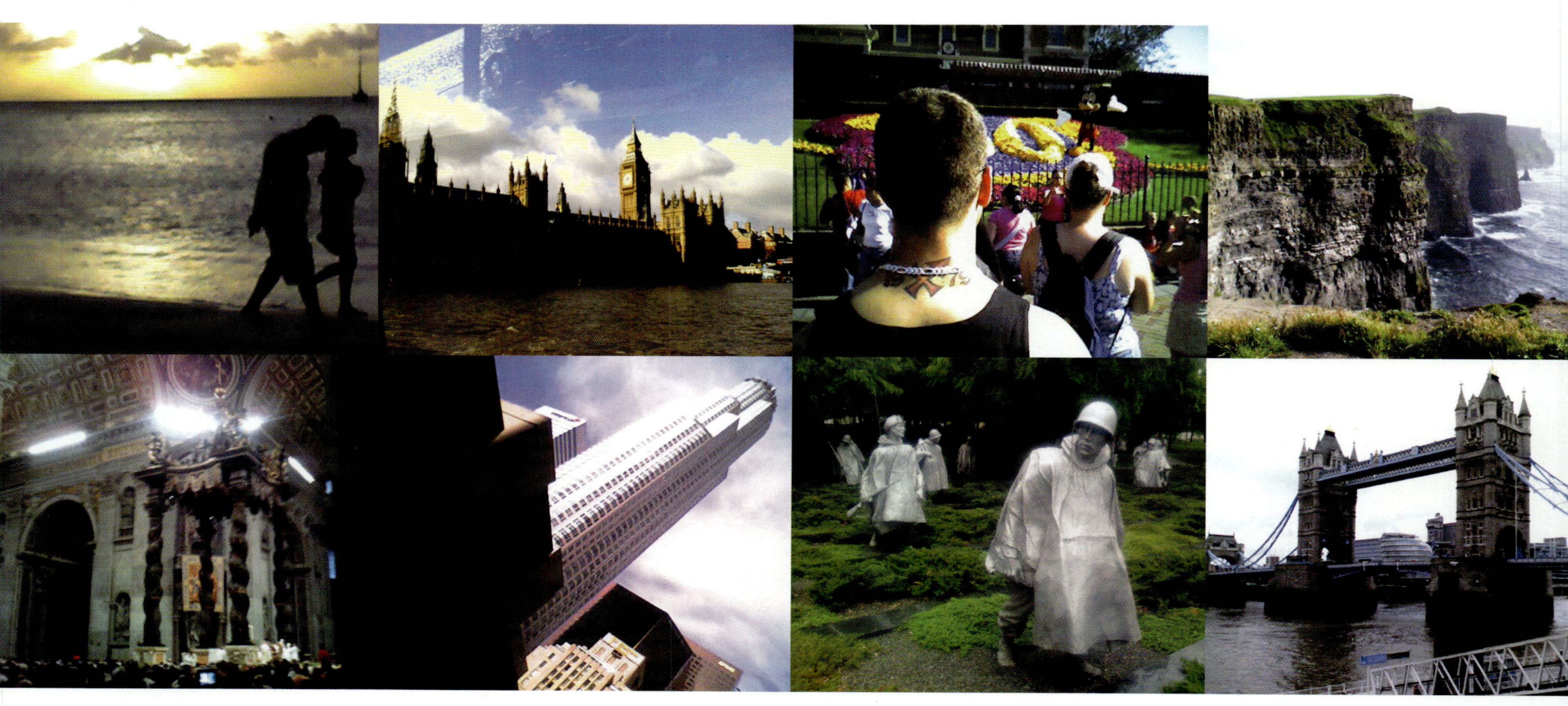

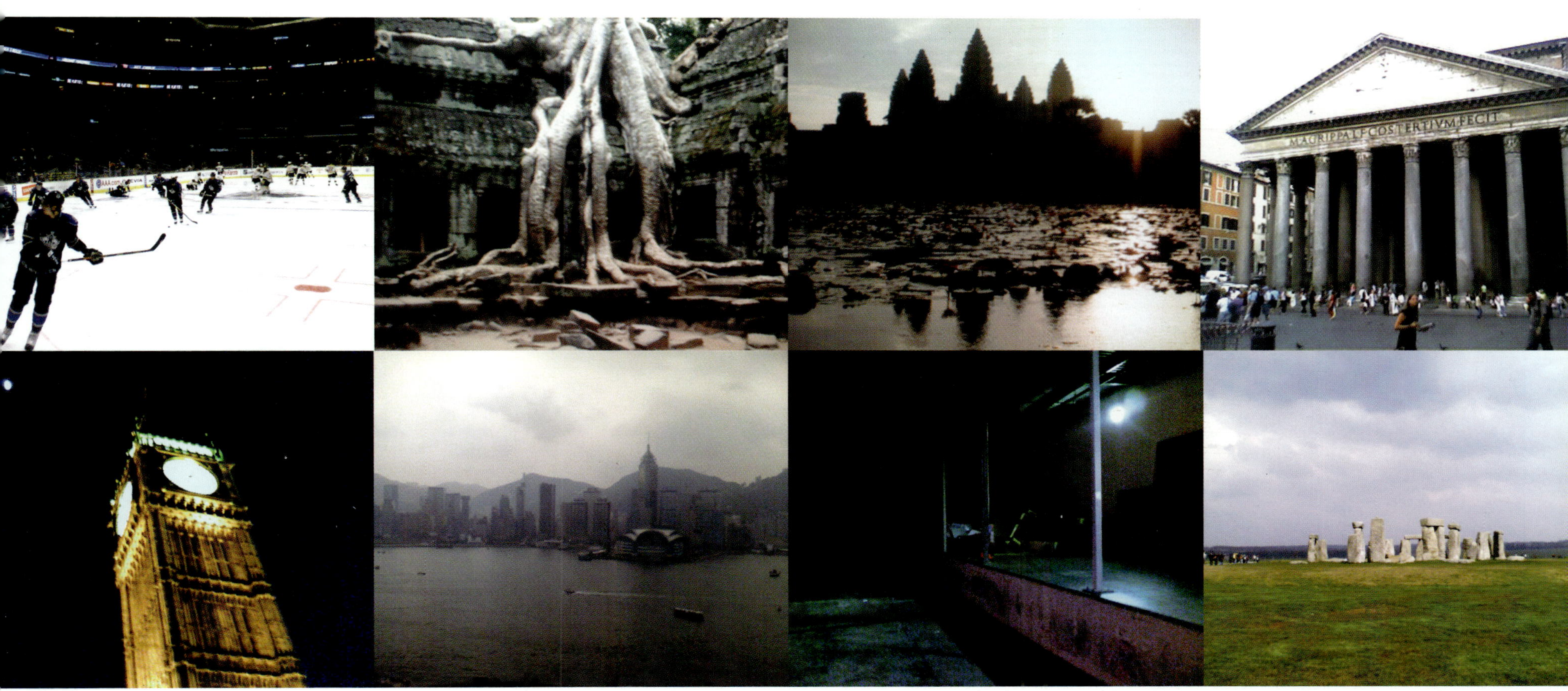

MAGRIPPALFCOSTERTIVMFECIT

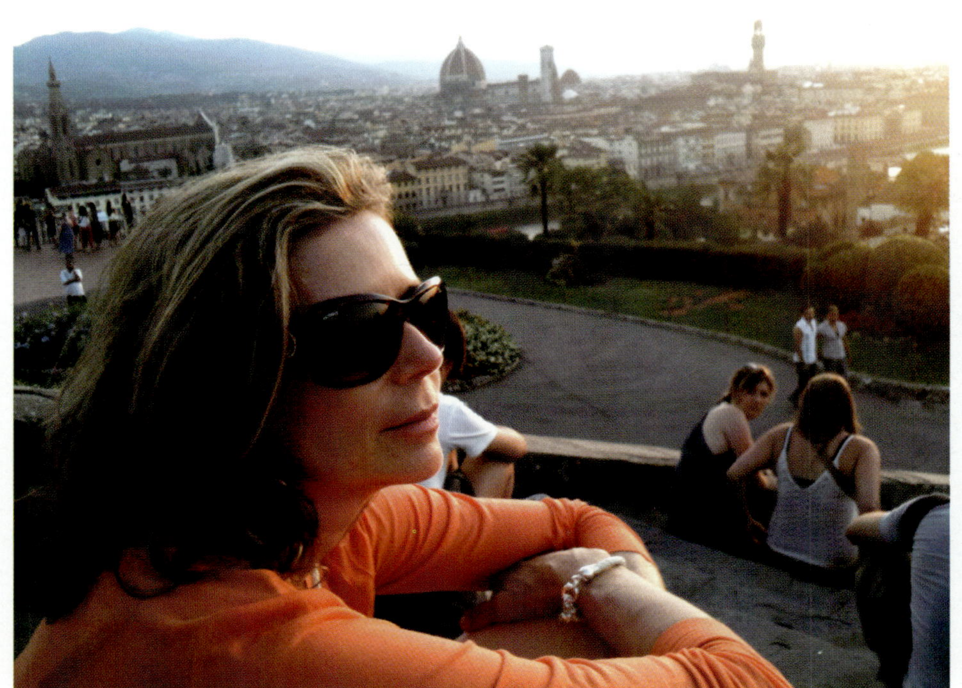
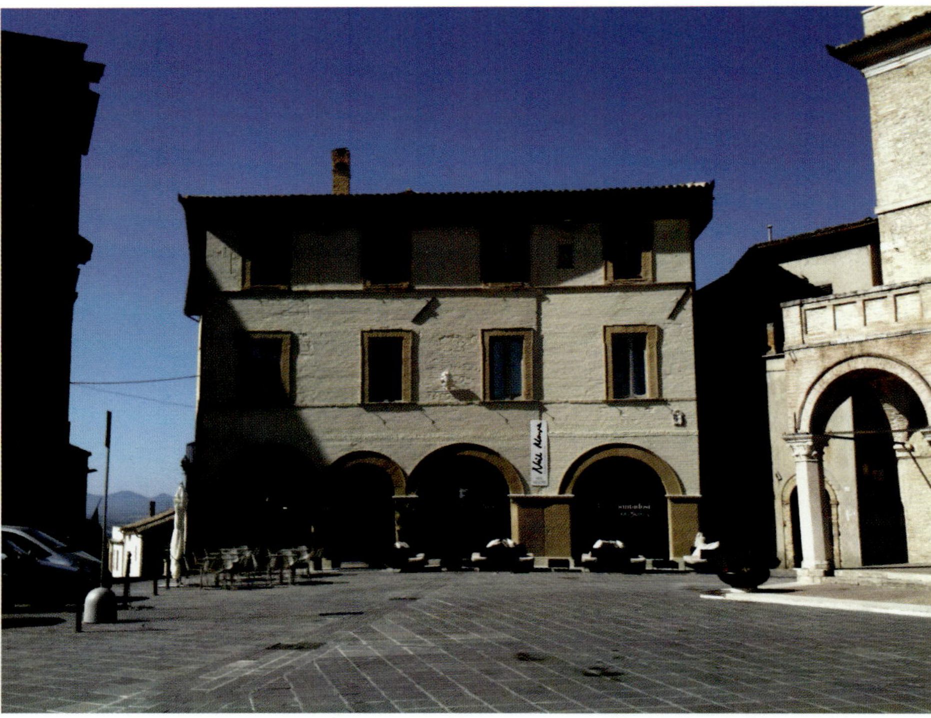
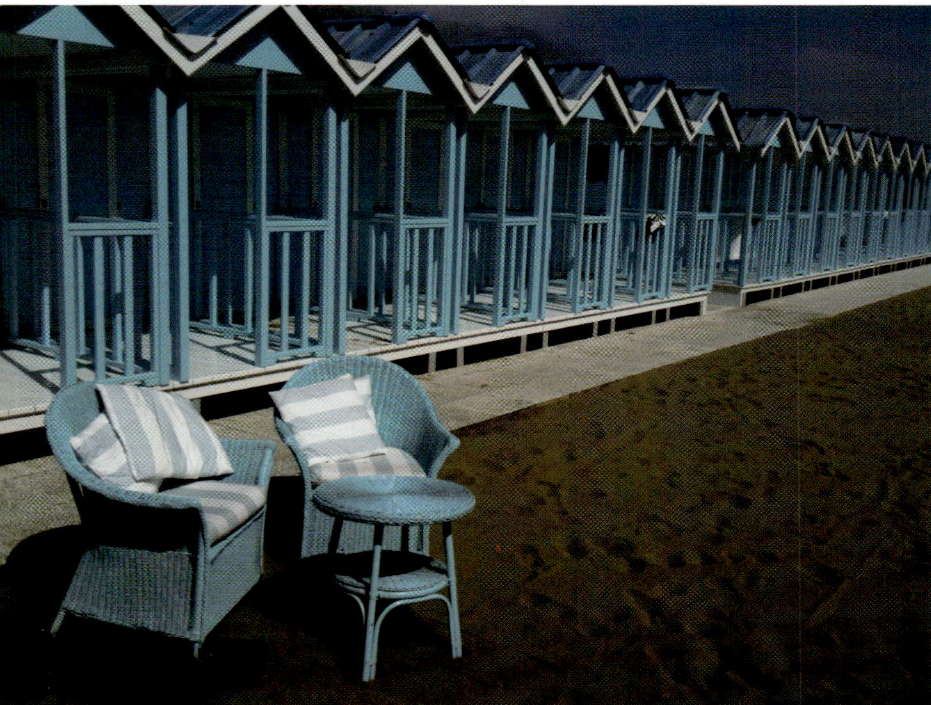
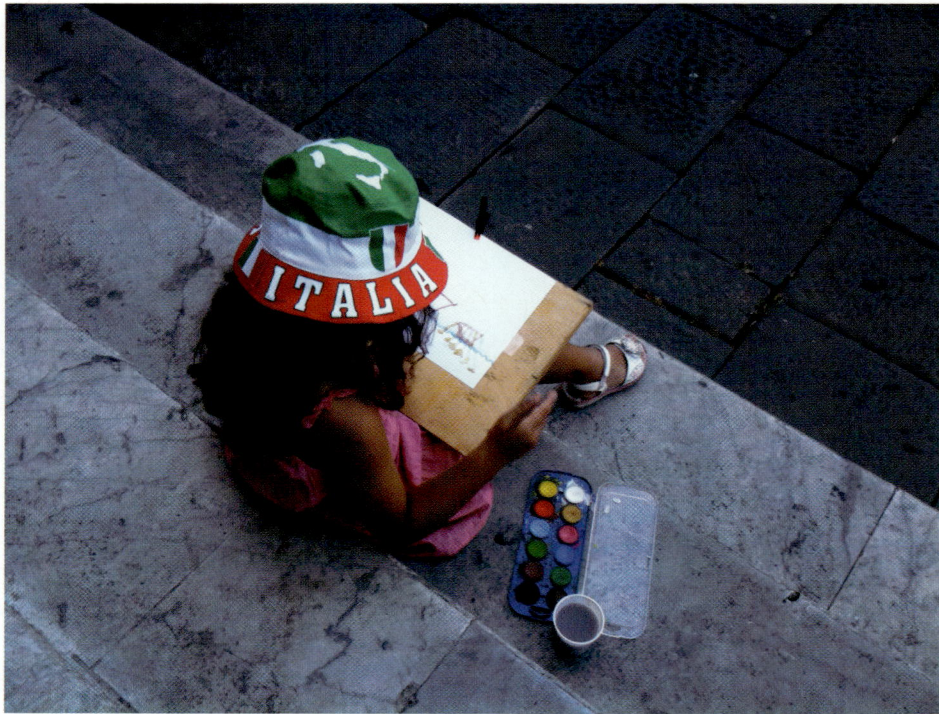

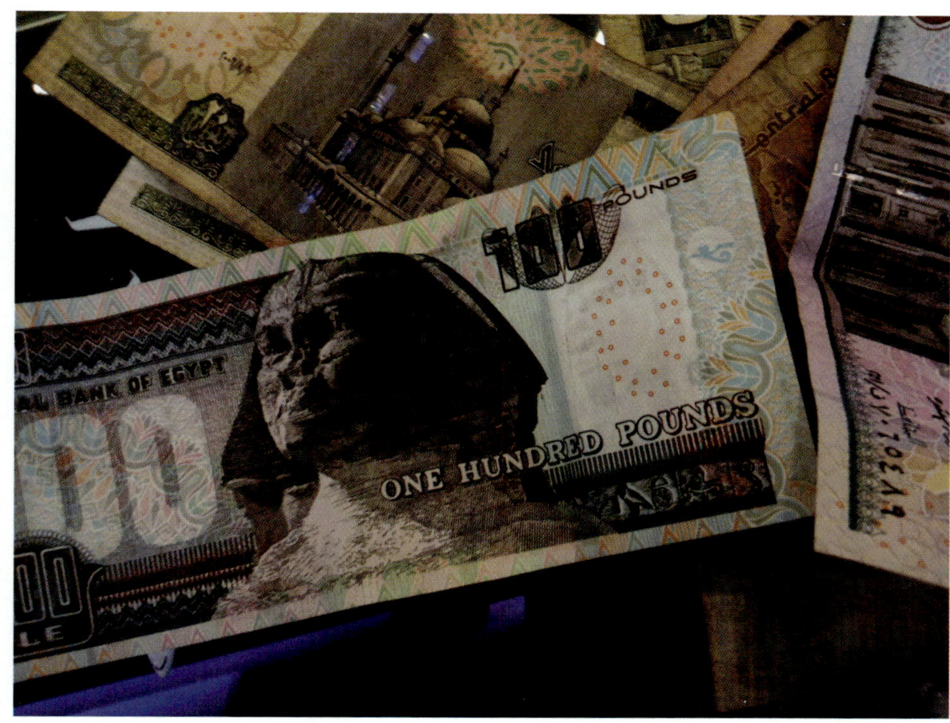
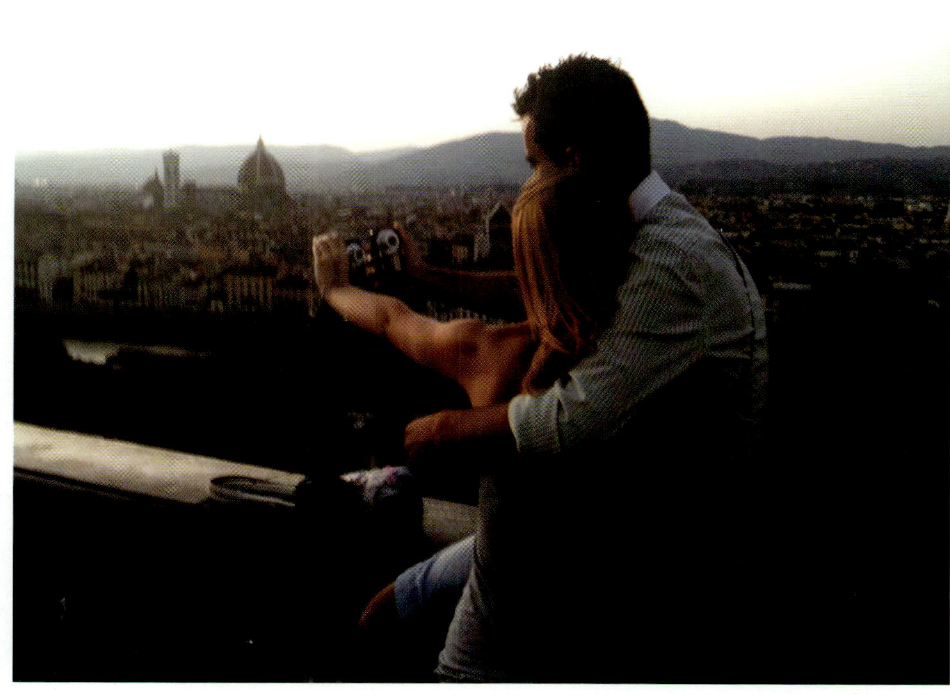
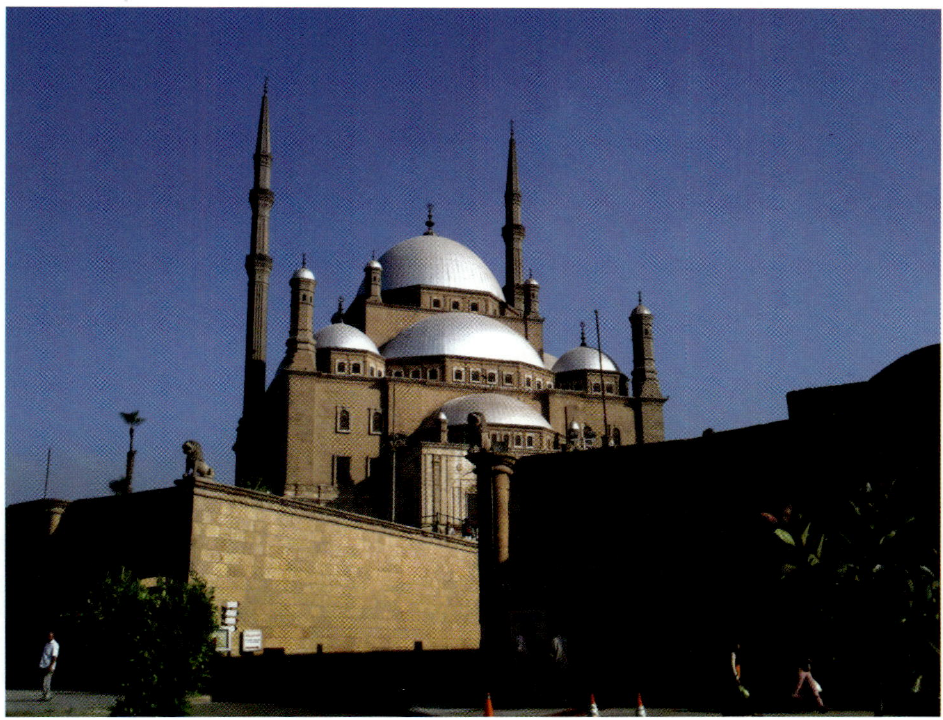
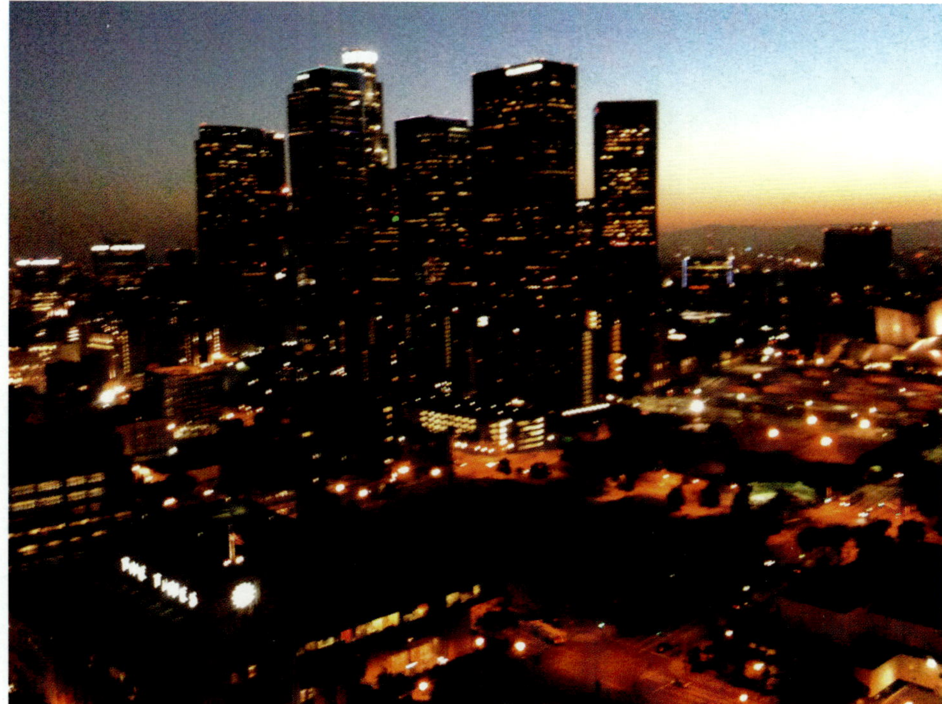

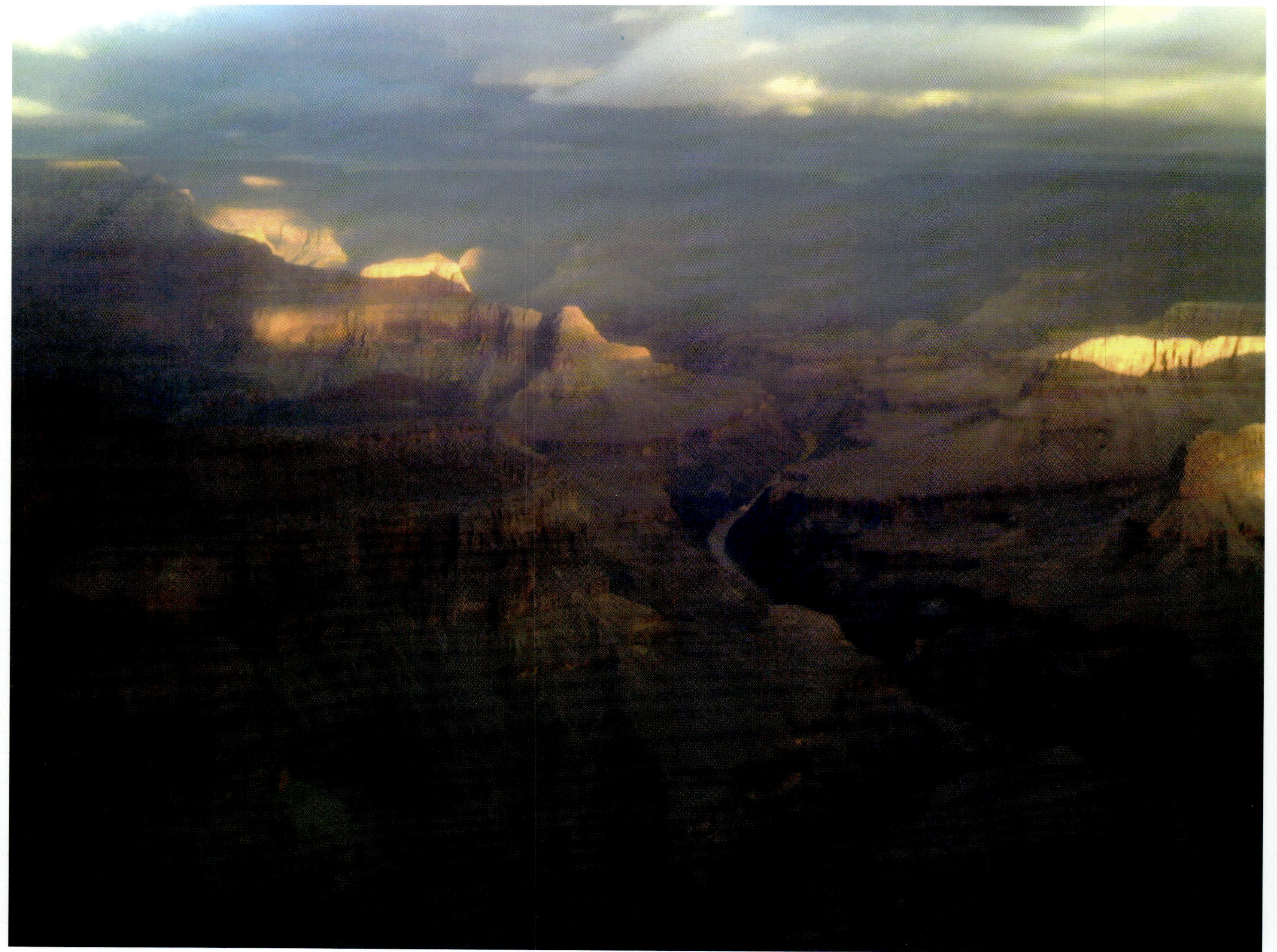

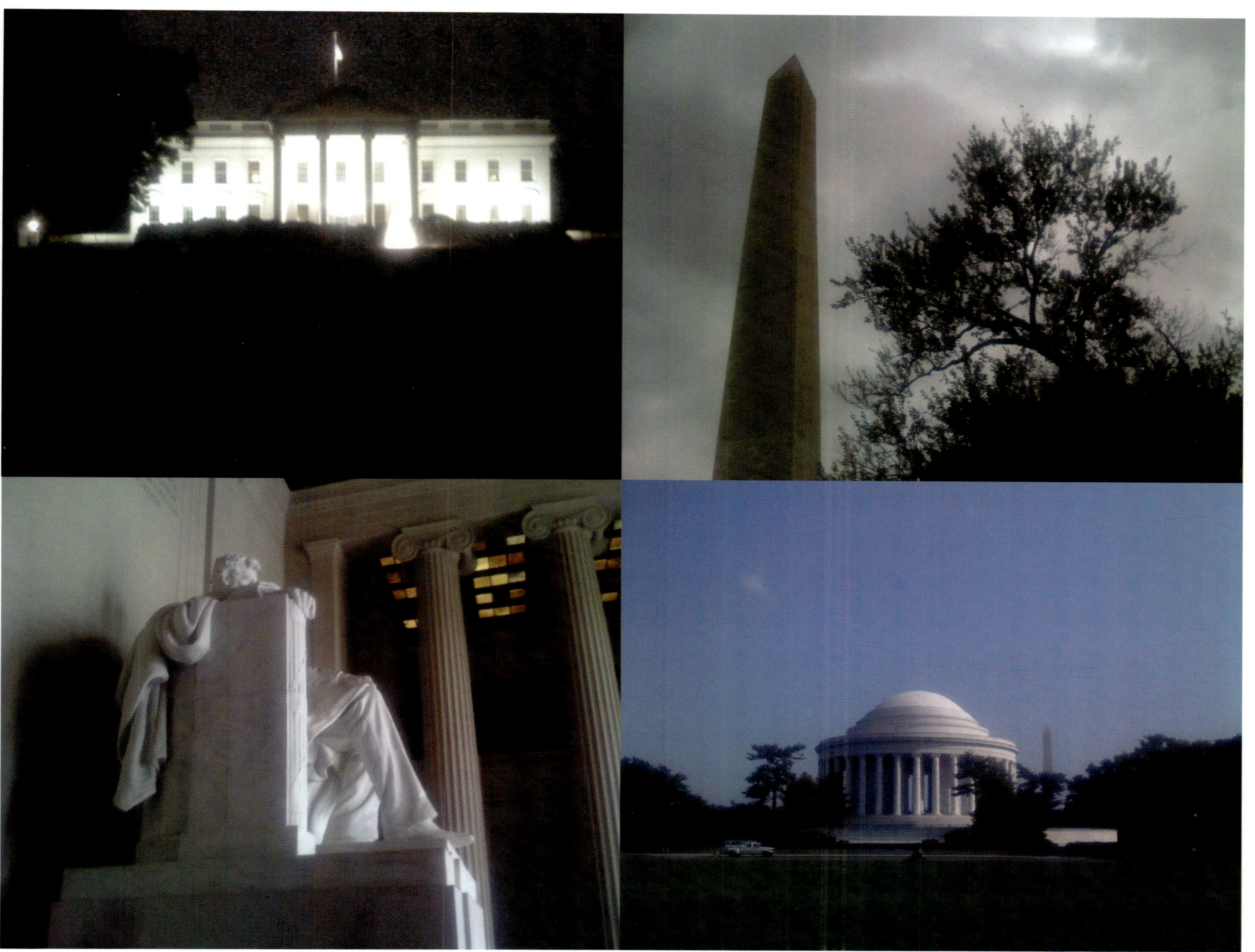

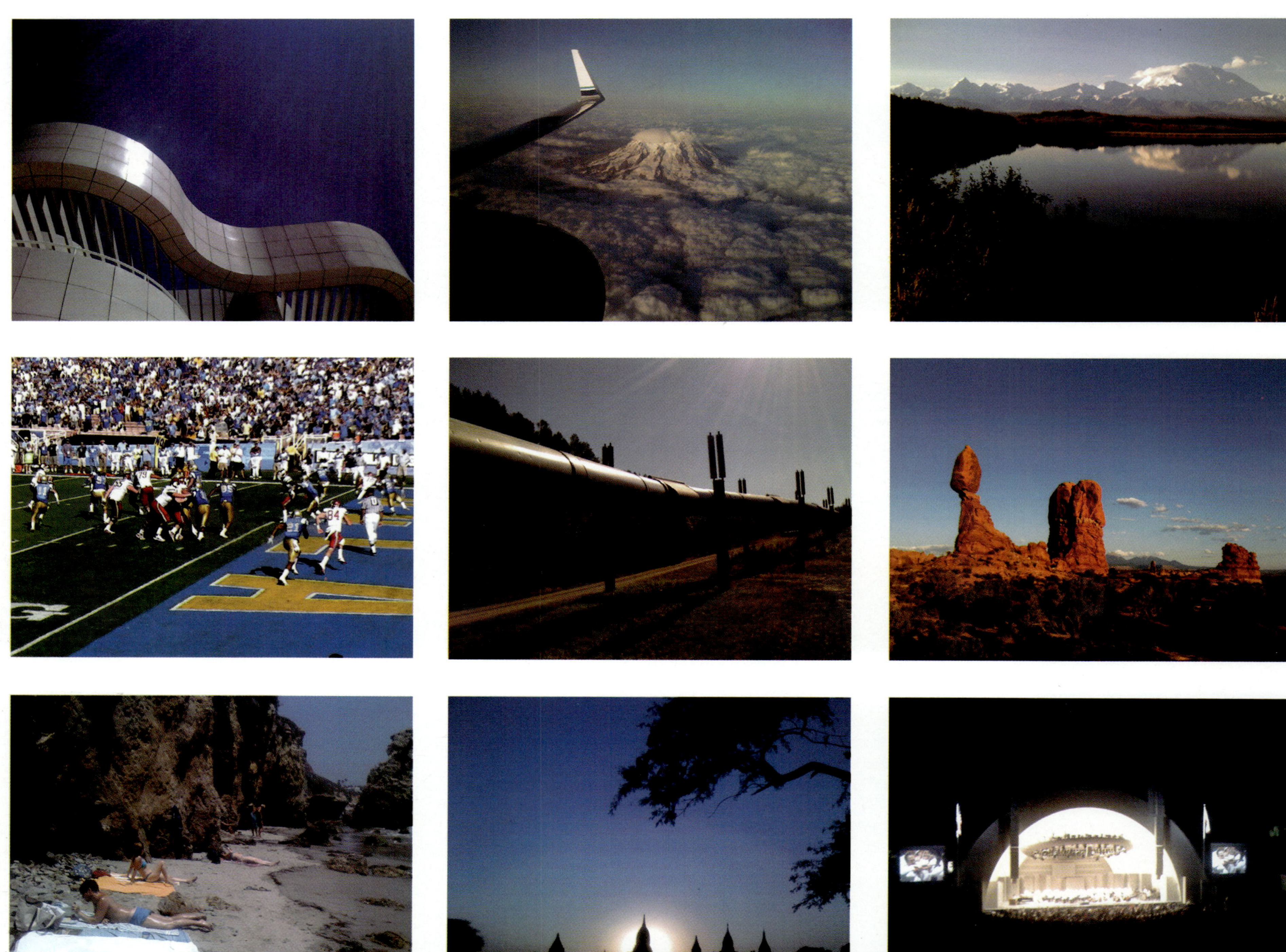

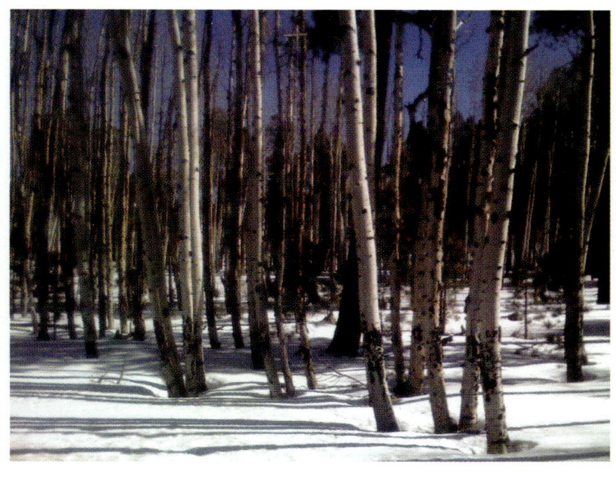
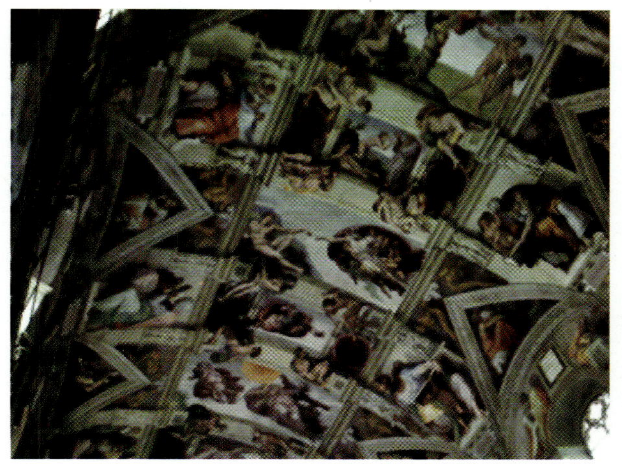
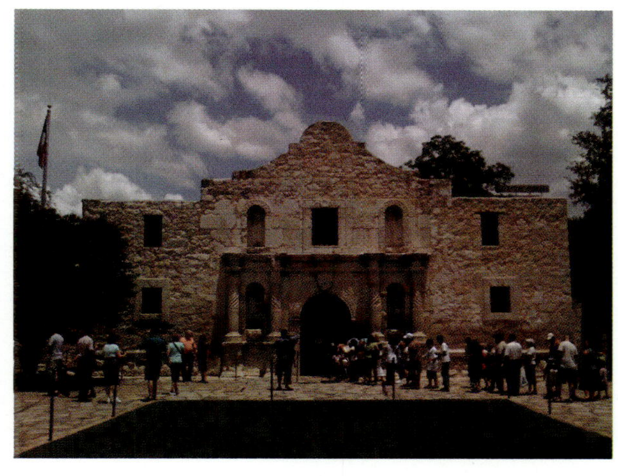
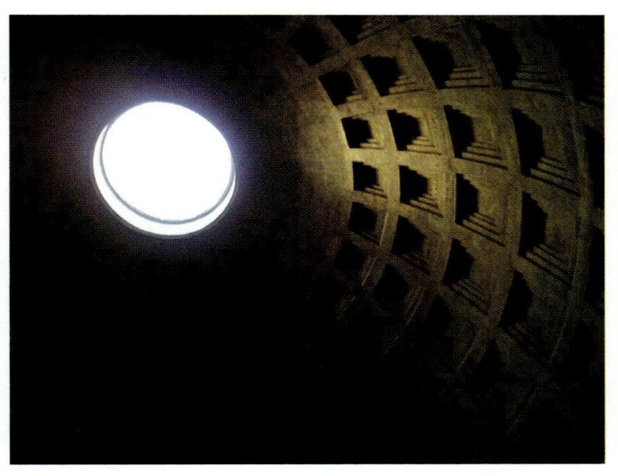
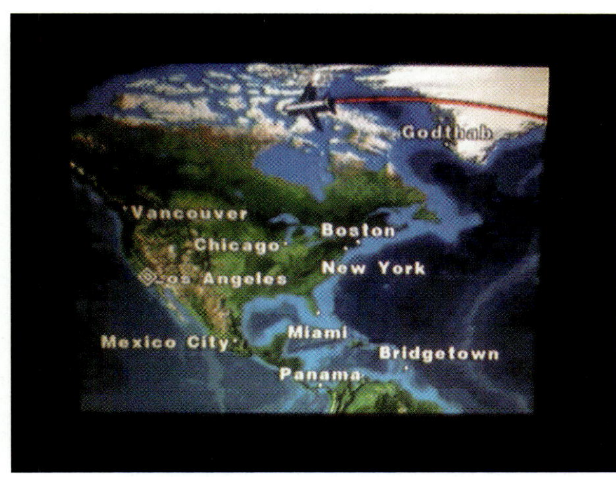
Godthab
Vancouver
Chicago
Boston
Los Angeles
New York
Mexico City
Miami
Bridgetown
Panama
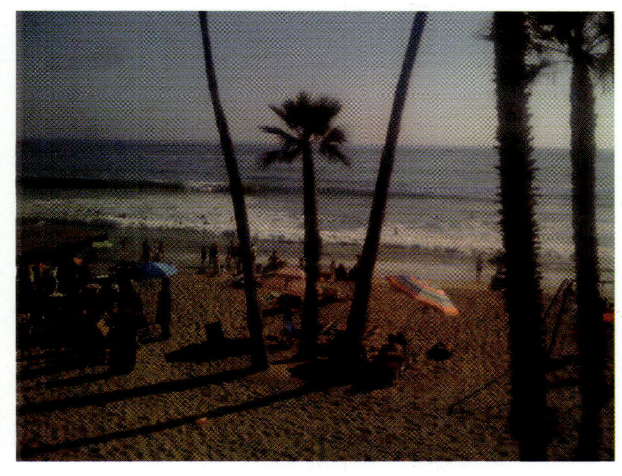
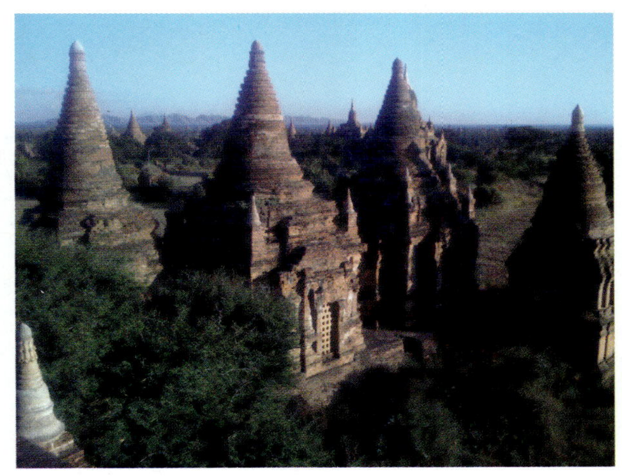
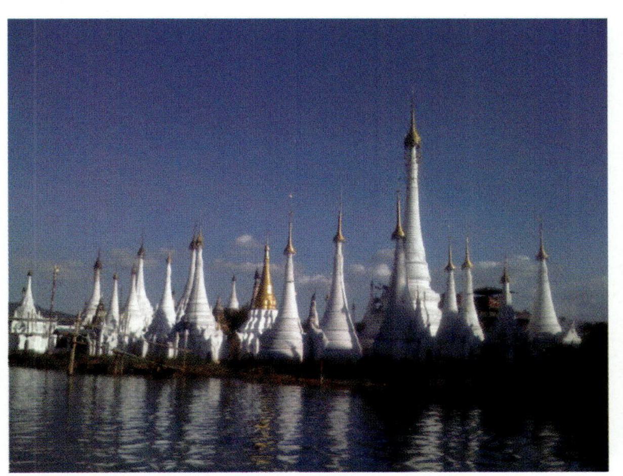
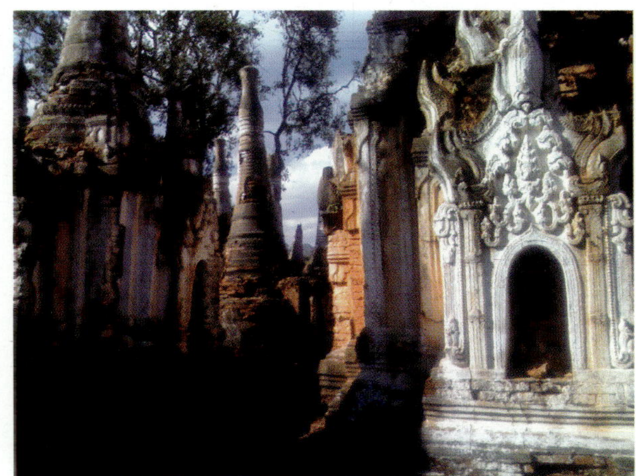

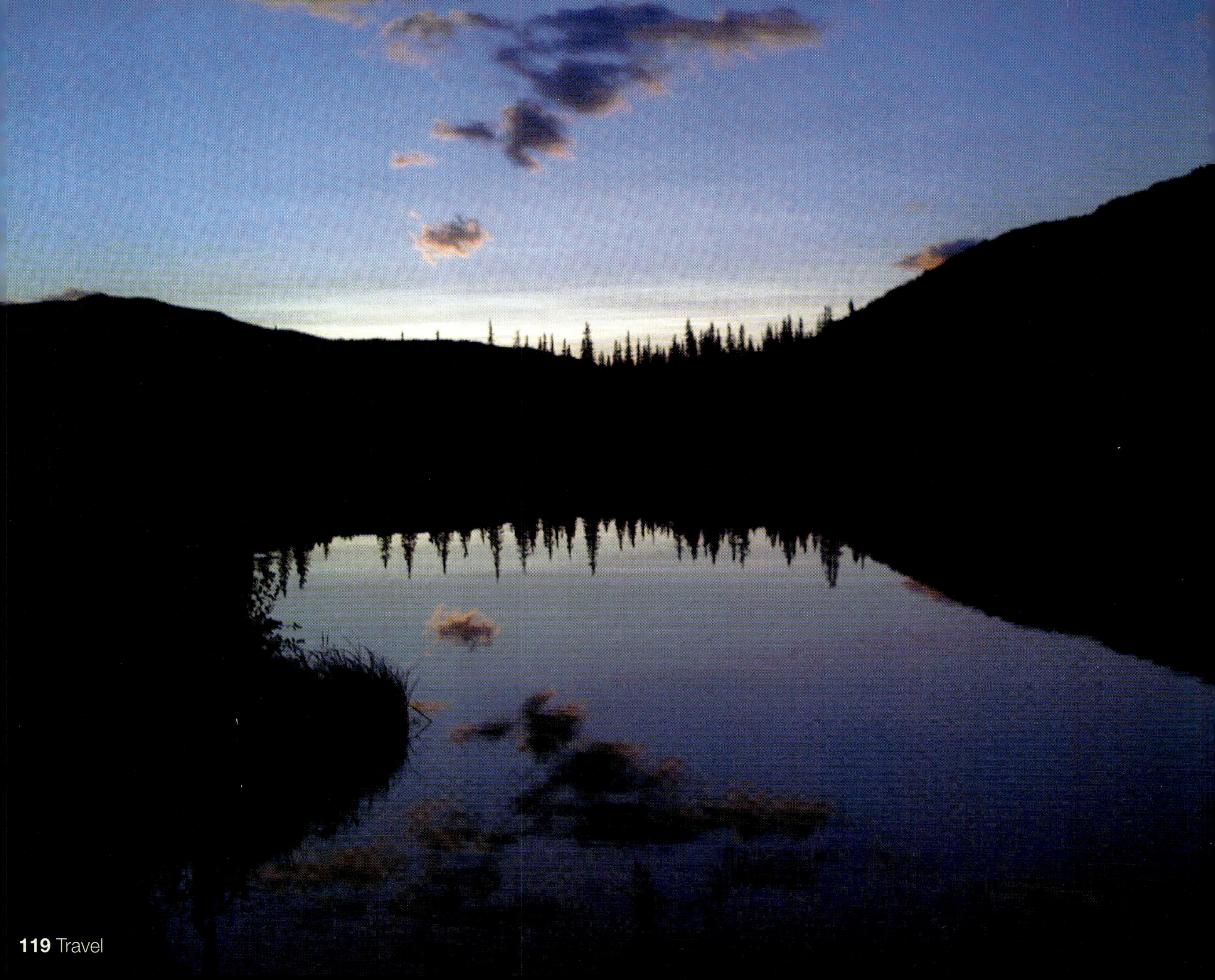

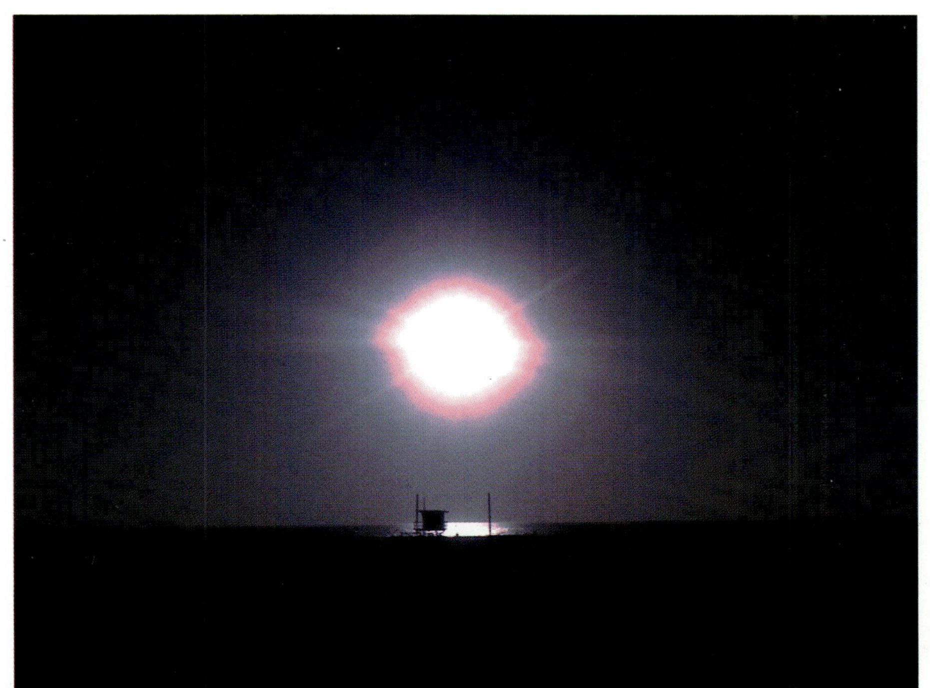

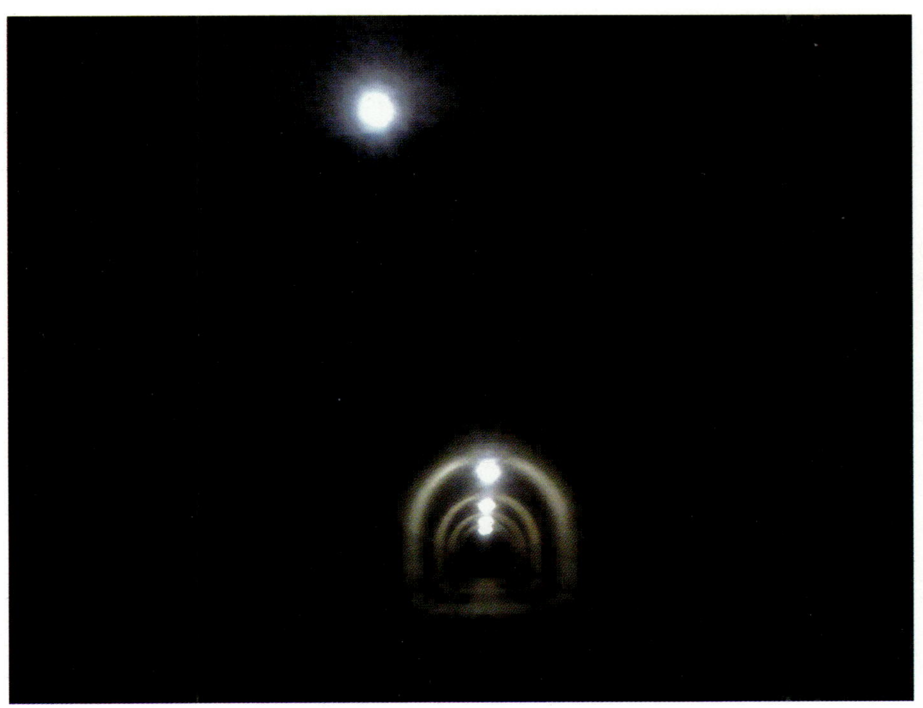

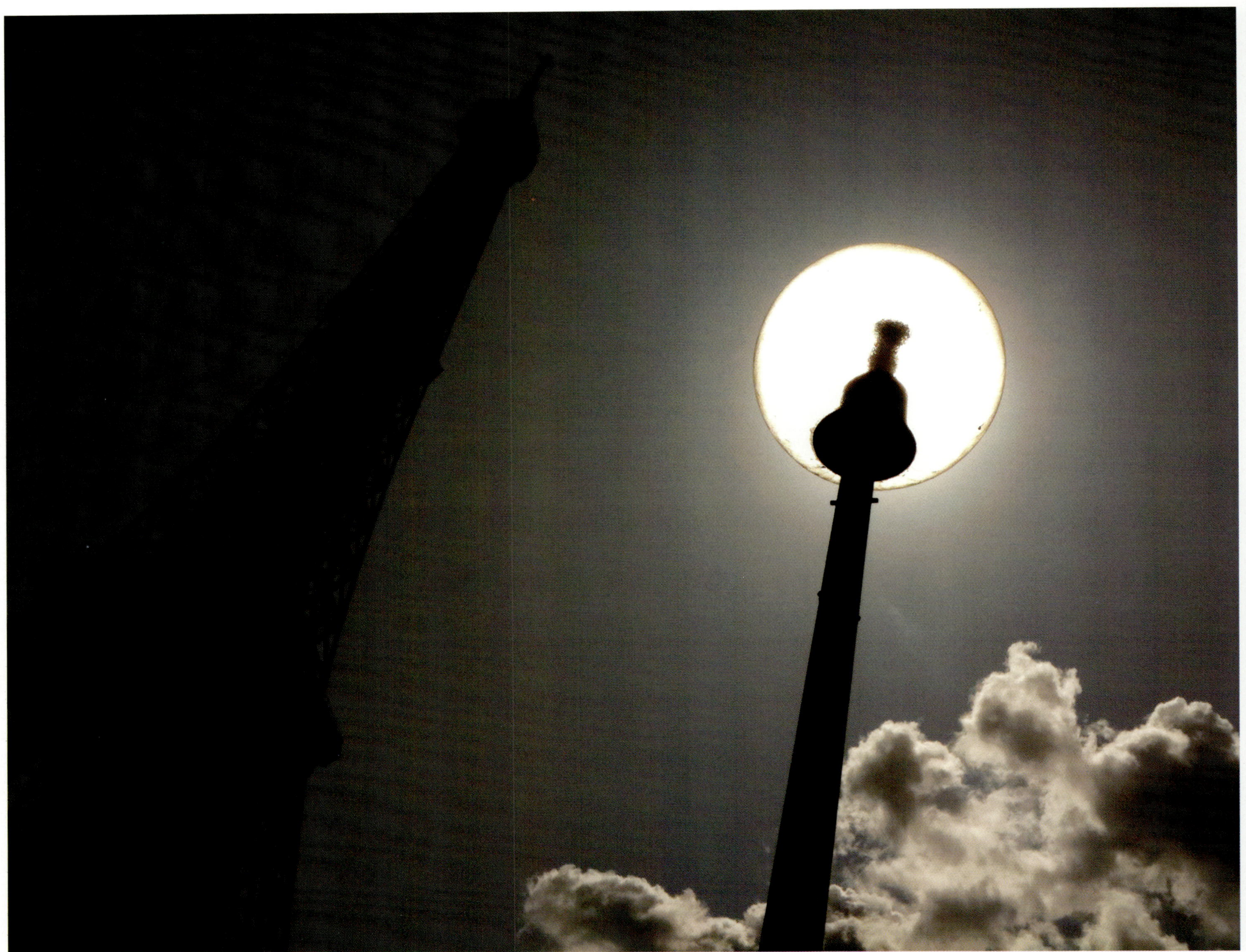

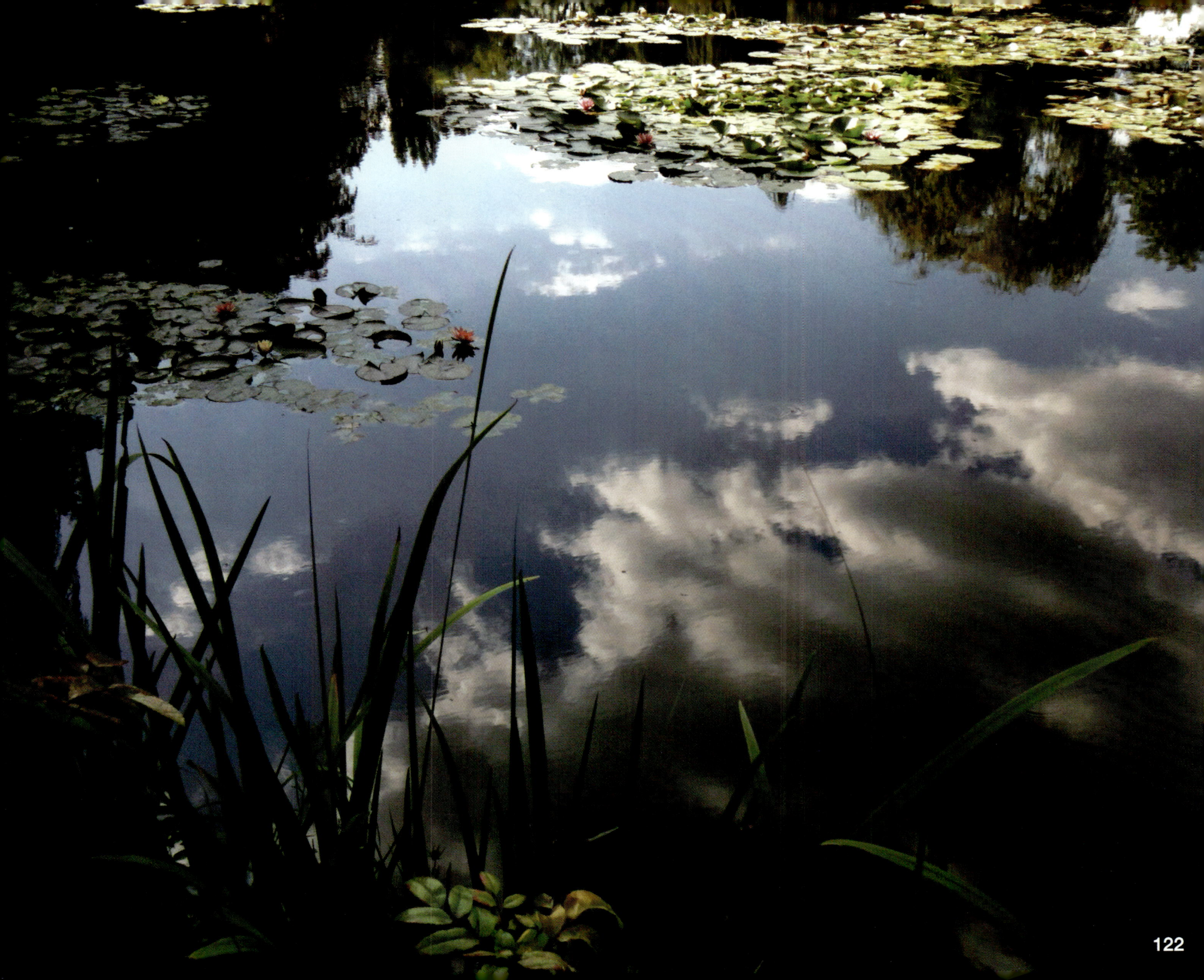

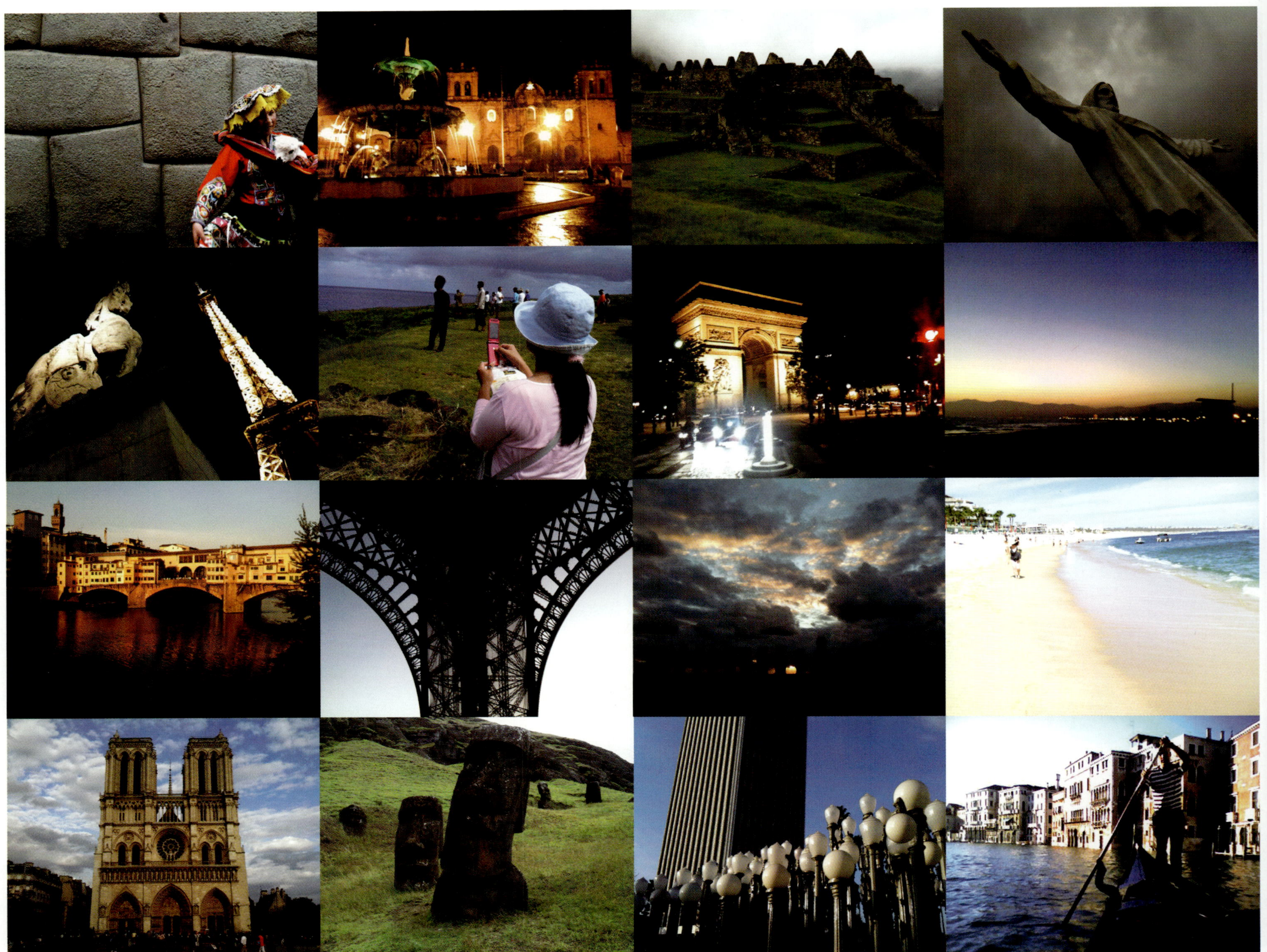

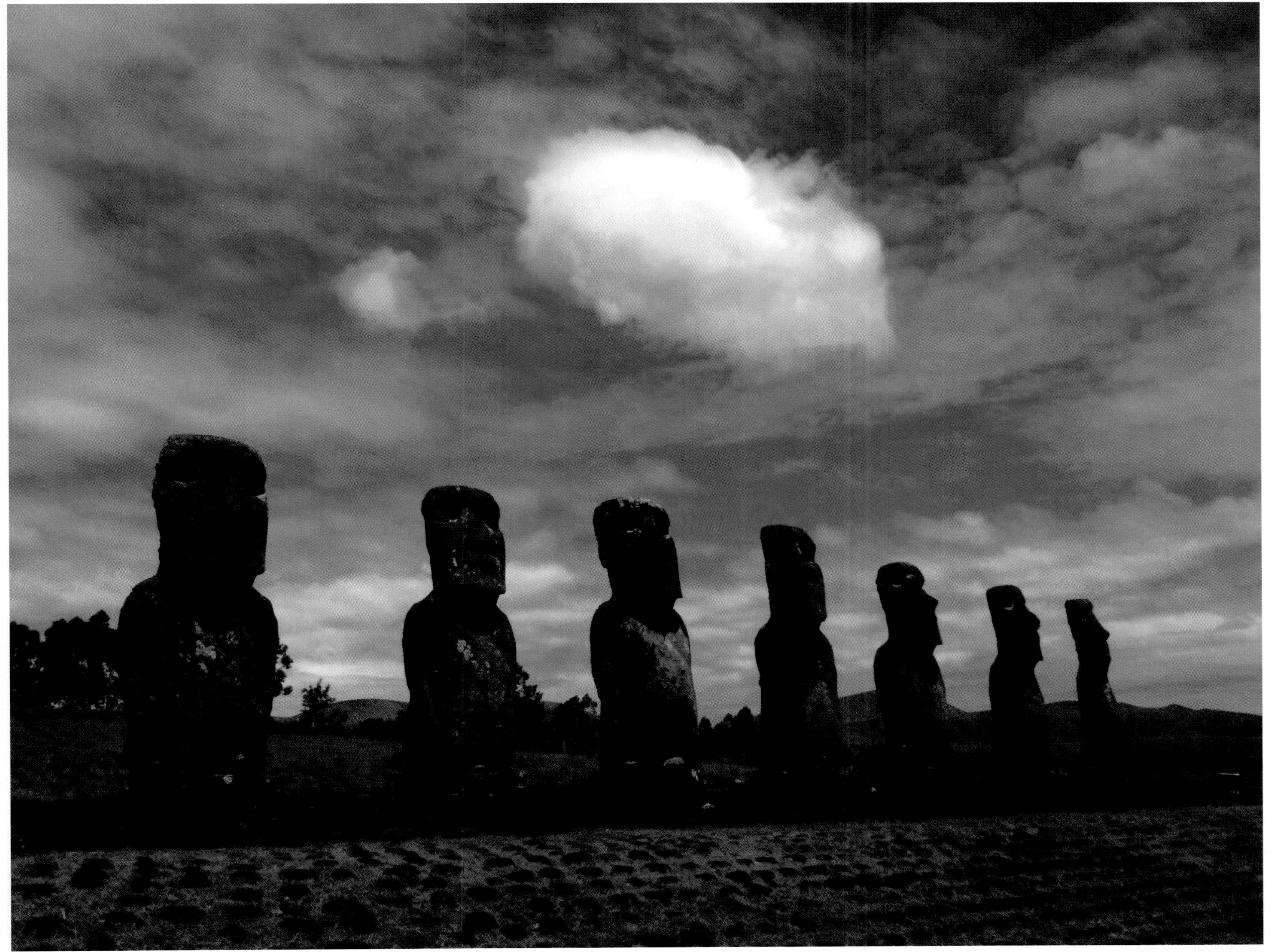

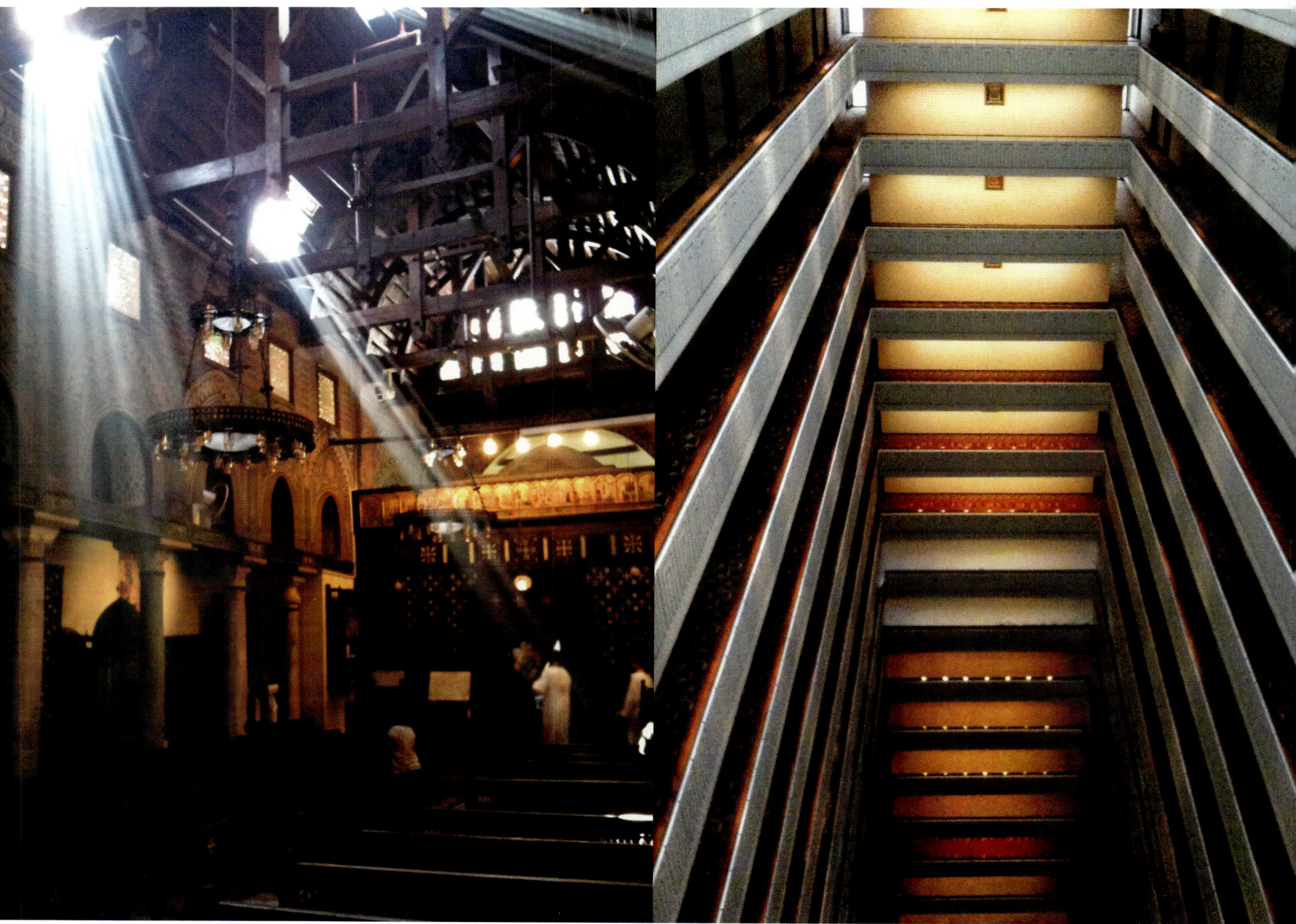

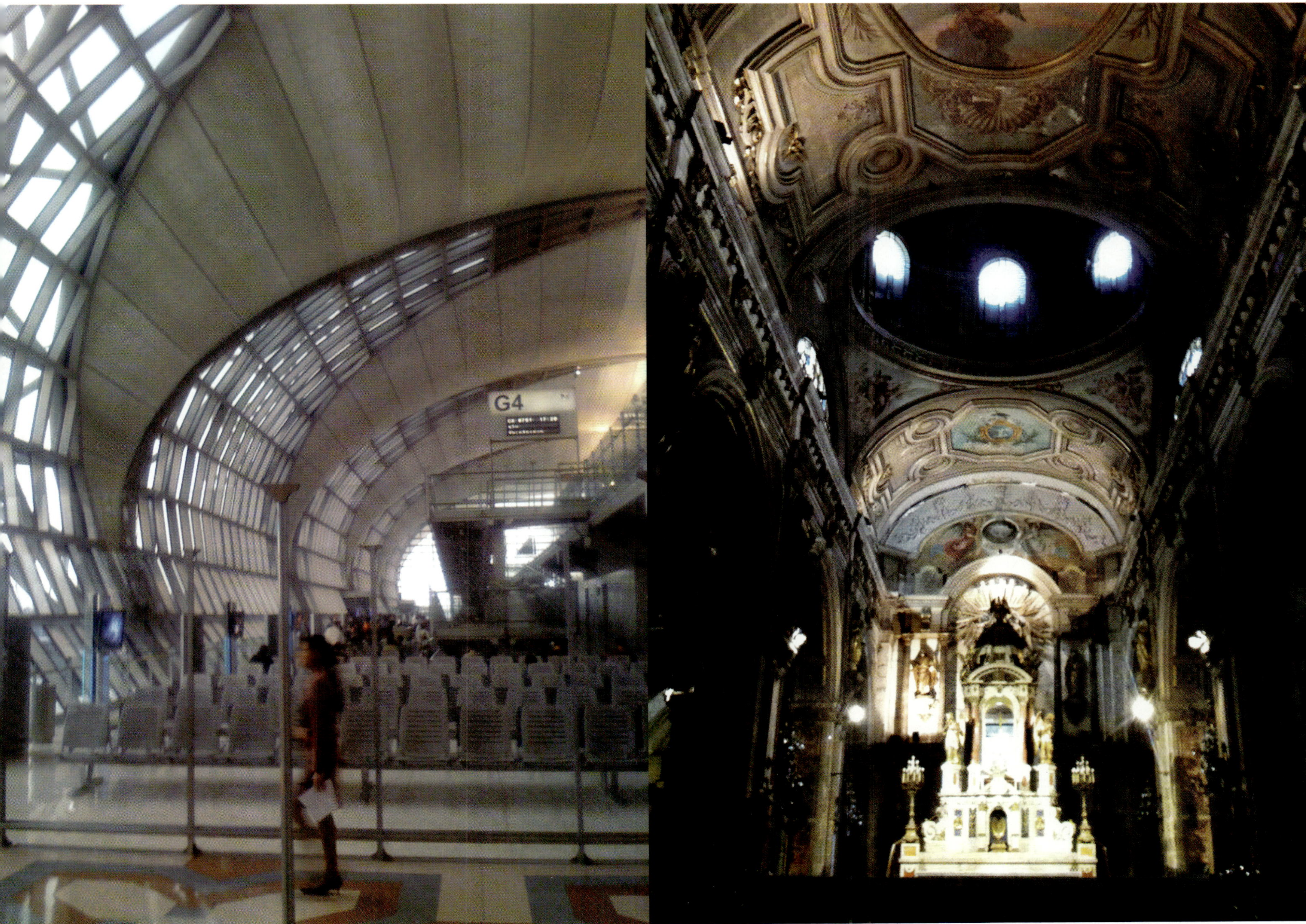

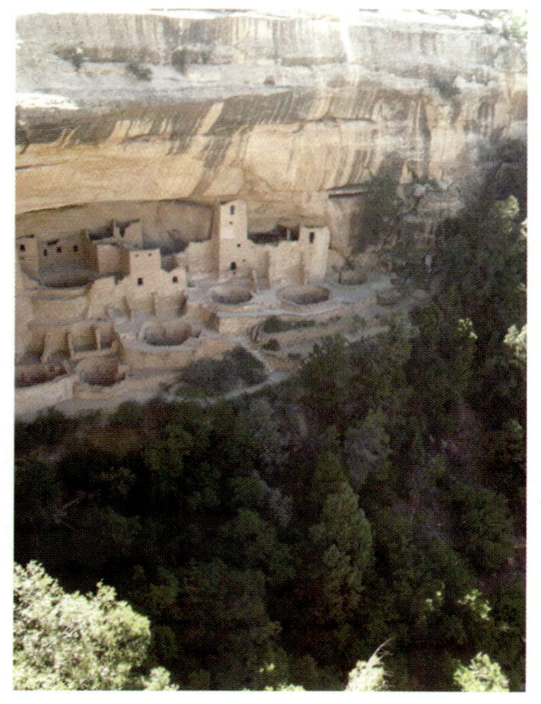
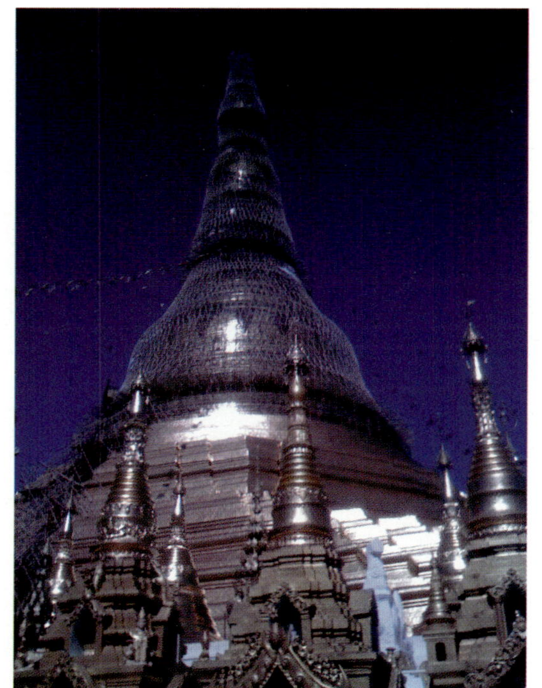
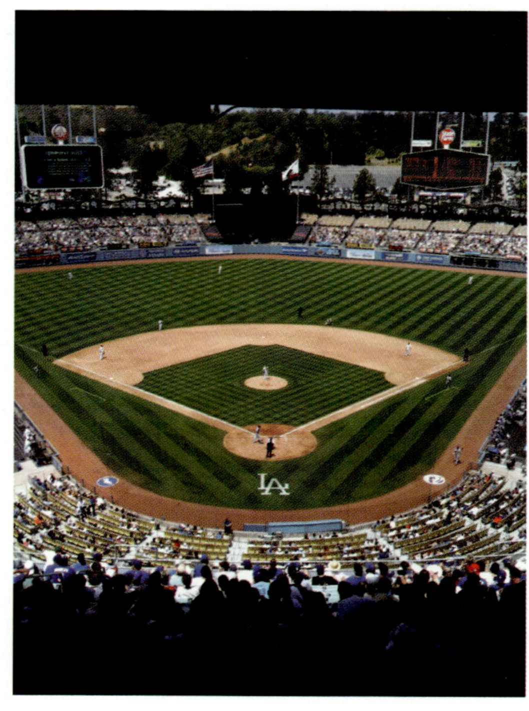
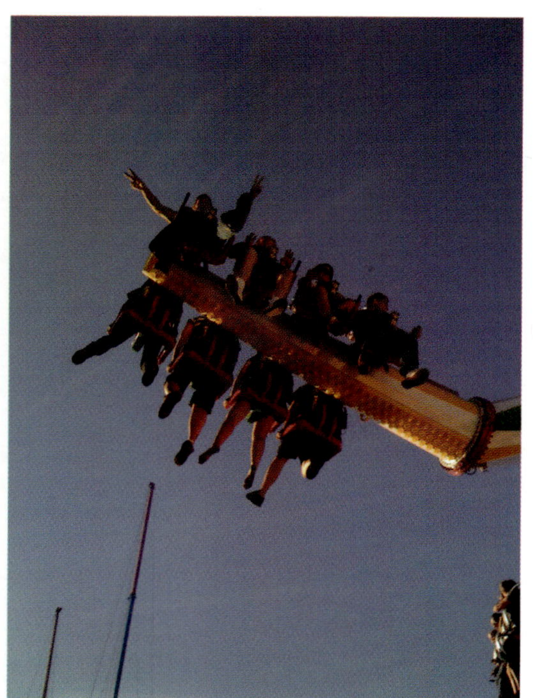
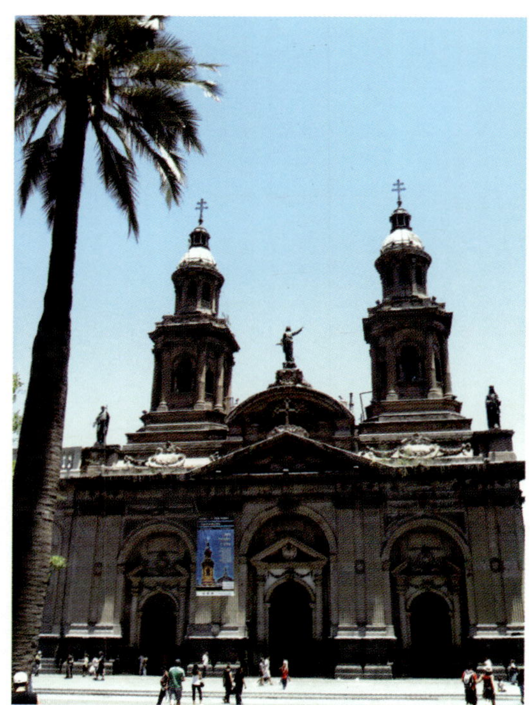
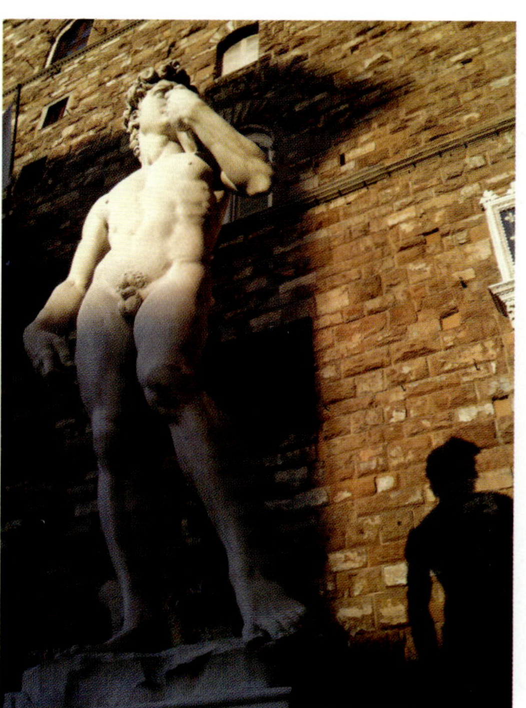

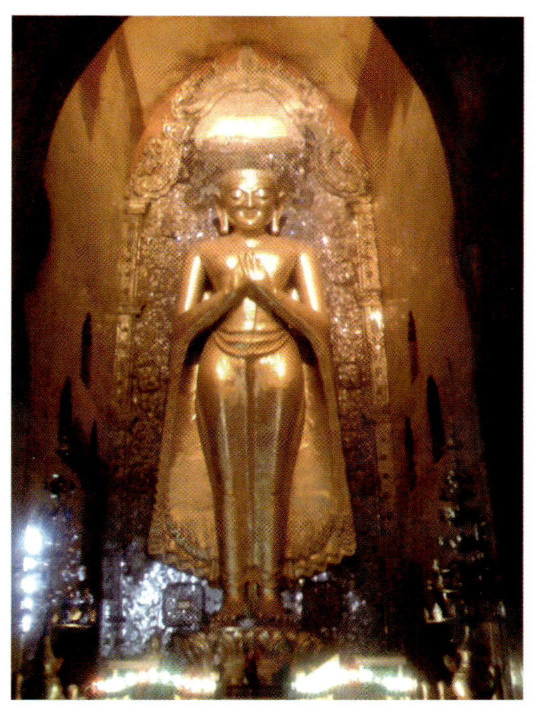
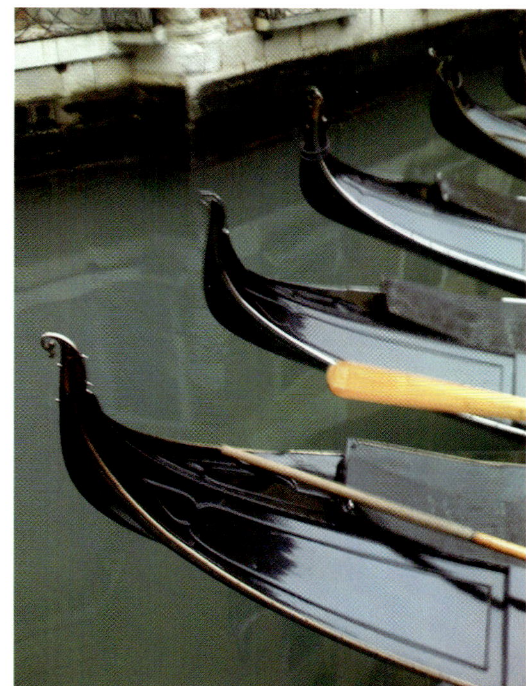
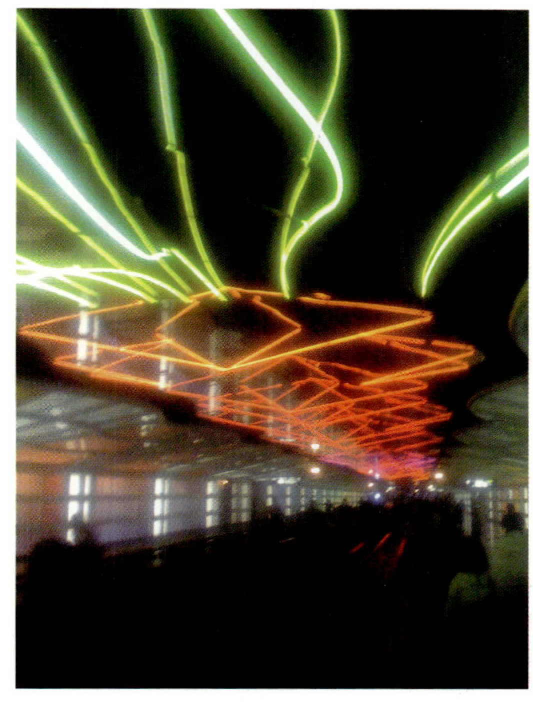
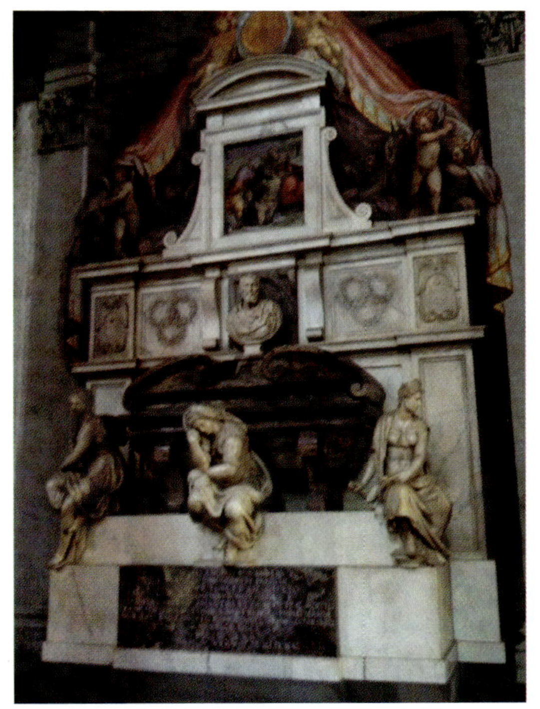
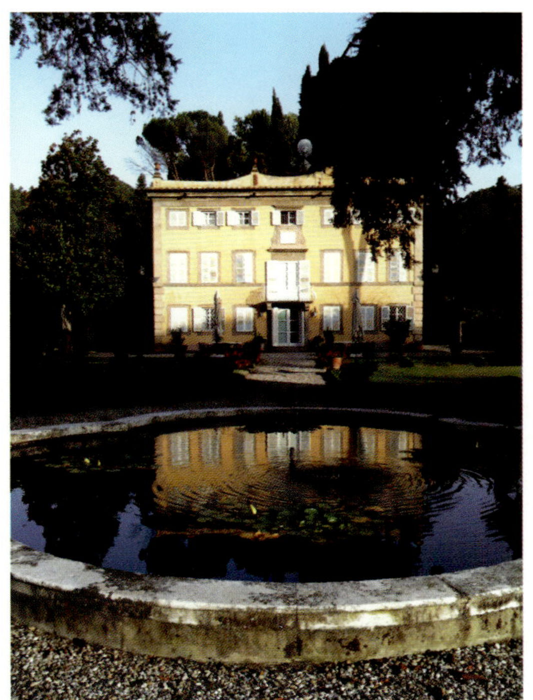
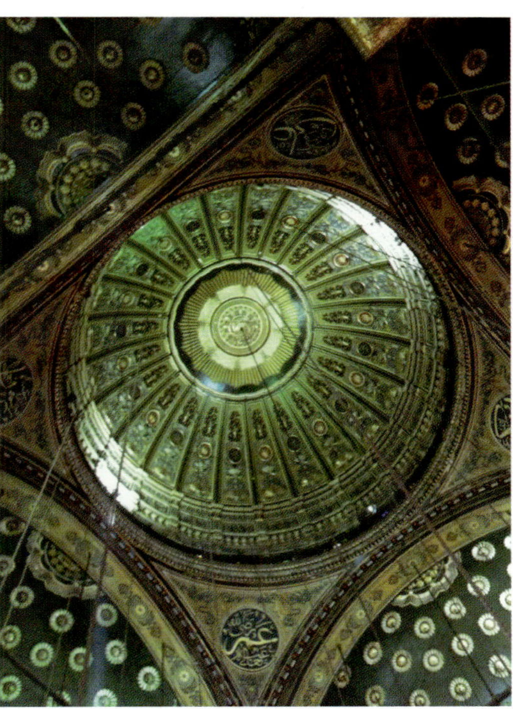

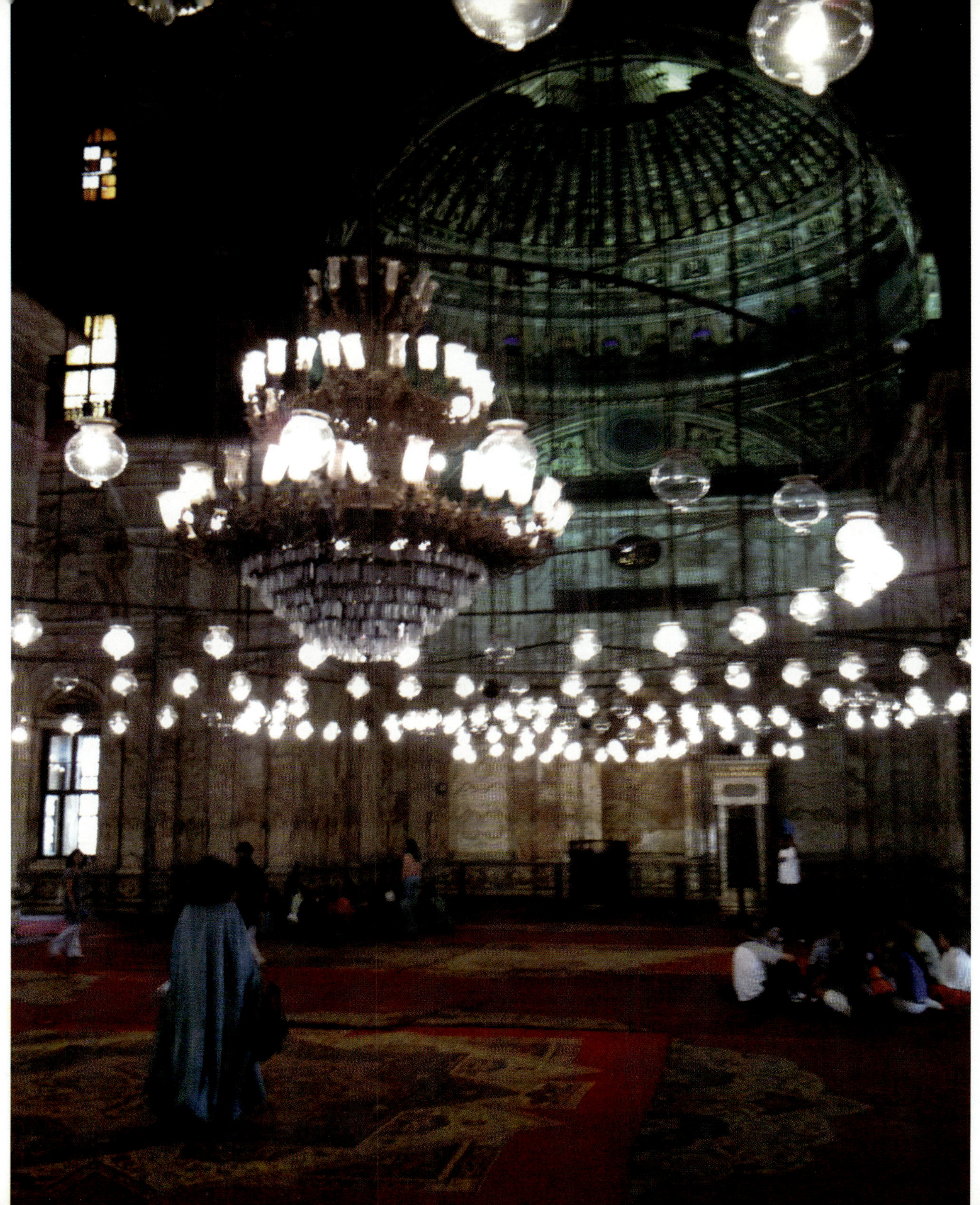

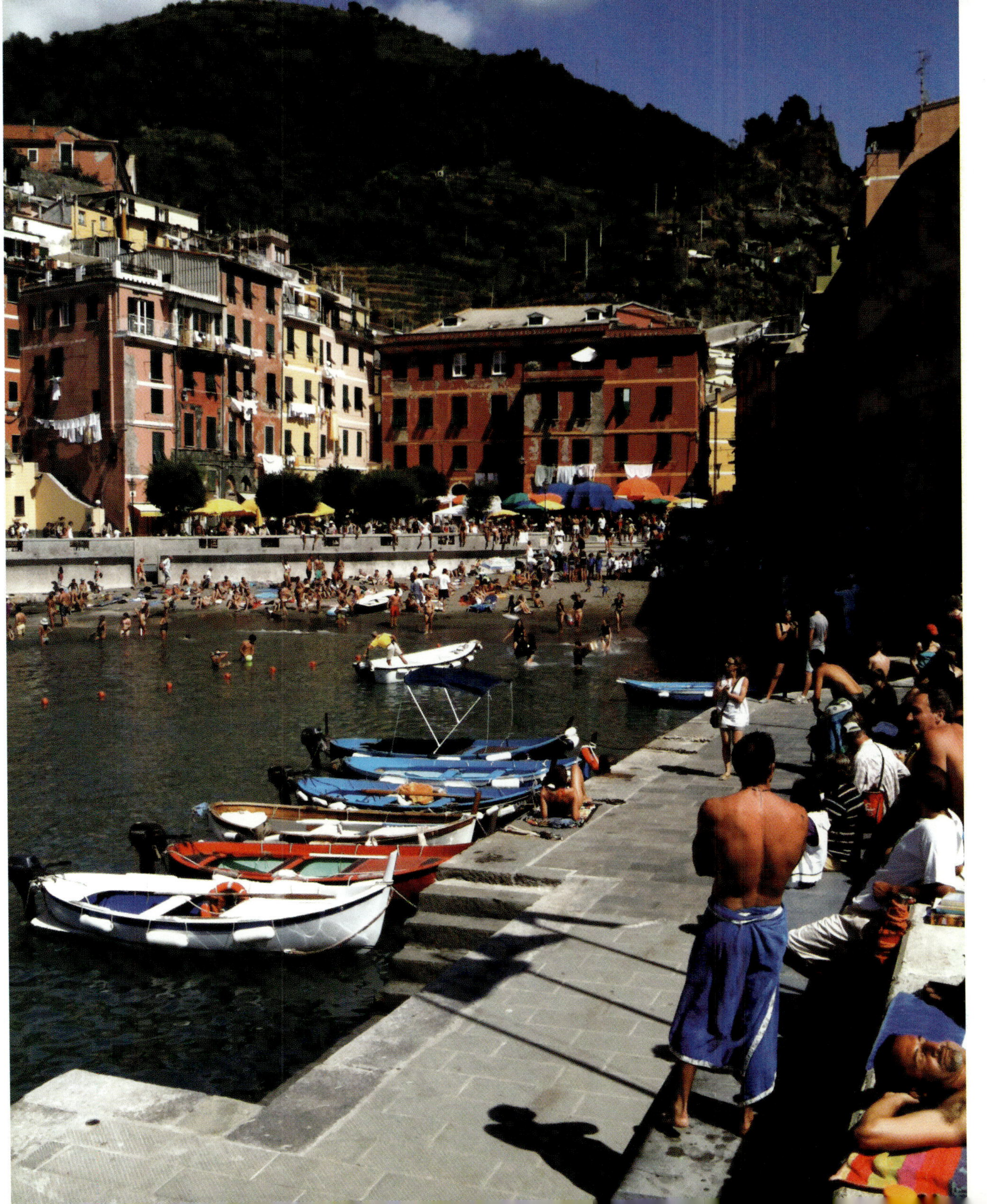

BIOGRAPHY

Voted by Frontier Magazine as one of Los Angeles' 25 most promising artists, celebrity and fashion photographer Cristopher Lapp has photographed some of the world's greatest celebrities – Meryl Streep, Sir Anthony Hopkins, Giorgio Armani, and Karl Lagerfeld, just to name a few. His work has appeared in Vogue, GQ, and "W" magazines. Cristopher's career began with hands-on training in Milan, working in front of and behind the camera for several well-known fashion photographers. In 2004, Cristopher returned to Italy for a post-baccalaureate degree in fashion photography at ADFF/CLICK UP. Apprenticeships with *National Geographic* and the *American Film Institute* were the precursors to Cristopher's extensive photographic coverage of all seven continents.

While regularly participating in solo and group exhibitions worldwide, Cristopher's fine art work has been shown at the annual Venice Art Walk, LA Art Mart, and has appeared at the Center for the Study of Popular Culture. His work can be seen on the walls of the Orange County Center for Contemporary Art and the Odyssey Theater. His fine art prints are collected by some of Hollywood's most elite, including Sir Elton John and the late Michael Jackson.

Cristopher is on the Board of the *American Society of Media Photographers – Los Angeles*, has taught photography at *The Braille Institute for the Blind* in Los Angeles, California and also volunteers his photographic services to *Camp Laurel*, a non-profit organization, serving children, youth and families living with HIV and AIDS.

Cristopher currently resides in Los Angeles, California.

www.CristopherLapp.com

Now it's your turn...

**Please visit www.InstantGratificationBook.com and click on the links
for Facebook or Twitter to upload your own mobile phone images.**